Waterpower in Lowell

JOHNS HOPKINS INTRODUCTORY STUDIES
IN THE HISTORY OF TECHNOLOGY

Waterpower in Lowell

*Engineering and Industry in
Nineteenth-Century America*

PATRICK M. MALONE

The Johns Hopkins University Press

Baltimore

The Johns Hopkins University Press
2715 North Charles Street
Baltimore, Maryland 21218-4363
www.press.jhu.edu

Library of Congress Cataloging-in-Publication Data

Malone, Patrick M.
    Waterpower in Lowell : engineering and industry in nineteenth-century America /
Patrick M. Malone.
        p. cm.
    Includes bibliographical references and index.
    ISBN-13: 978-0-8018-9305-6 (hardcover : alk. paper)
    ISBN-10: 0-8018-9305-4 (hardcover : alk. paper)
    ISBN-13: 978-0-8018-9306-3 (pbk. : alk. paper)
    ISBN-10: 0-8018-9306-2 (pbk. : alk. paper)
    1. Merrimack River (N.H. and Mass.)—Power utilization—History—19th century.
2. Water-power—Massachusetts—Lowell—History—19th century. 3. Hydraulic
engineering—Massachusetts—Lowell—History—19th century. 4. Merrimack River
(N.H. and Mass.)—Power utilization—History—19th century. I. Title.
    TC425.M4M35 2009
    621.209744'409034—dc22   2008052450

A catalog record for this book is available from the British Library.

*Special discounts are available for bulk purchases of this book. For more information,
please contact Special Sales at 410-516-6936 or specialsales@press.jhu.edu.*

For Charles Parrott,
designer of waterwheels
and defender of the Lowell Canal System

Contents

....

Acknowledgments

···

This project began in 1971, when the University of Pennsylvania funded a summer of research and data collection for a graduate seminar that I planned to offer on Lowell. The students in that seminar, particularly Larry Lankton, who focused on the power canals of Lowell, were the first of many to assist in the collaborative scholarship that made this book possible. Louis Eno, Lewis Karabatsos, John Goodwin, and Gordon Marker welcomed me and my students to Lowell and connected us to its ardent preservation community. The Historic American Engineering Record (HAER) asked me to co-direct the recording of the Lowell Canal System in 1974–75, working with a talented team that included supervisory architect Chuck Parrott and historian of technology Charlie Hyde. Mel Lezberg, the modern equivalent of James B. Francis, provided us with office space in the Boott Cotton Mills and full access to the notebooks, maps, and drawings of Locks & Canals. Brown University gave me sabbatical leaves for writing and research grants for illustrations. The Dibner Institute at MIT twice provided me with senior fellowships. I am very grateful for feedback from fellows, staff, and visitors at the Dibner, where I made presentations and wrote articles on Lowell. Pat Martin, editor of *IA: The Journal of the Society for Industrial Archeology*, gave permission to use pieces from two of those articles in this book.

My greatest debt is to Chuck Parrott, who has worked with me on Lowell for thirty-five years. He generously allowed me to incorporate parts of an *IA* article that we coauthored, loaned me his personal copy of *Lowell Hydraulic Experiments,* produced two illustrations for the book, and provided detailed comments on the entire manuscript. Amy Kendall enhanced historical images and added her own drawings for both figures and sidebars. Steve Lubar, Lyn Malone, and Terry Reynolds read every page and made astute suggestions. Gray Fitzsimons, Kerry Gagnon, Rick Greenwood, Robert Gordon, Mel Lezberg, and Martha Mayo reviewed sections at my request. For decades, Martha has provided an inexhaustible flow of expert opinion and key documents. Gray contributed copies of newspaper articles from his own research as historian of the Lowell National Historical Park. Claiborne Walthall gave me the note cards from his thesis project on Lowell and tabulated reams of statistics. Al Lorenzo found historical documents for me as he wrote his own studies of the canal system. Beth Belanger did research at my request and converted some of my earlier writings to digital format. Marti Frank offered perceptive ideas about steam power and co-generation. Larry Gross sent information on the Boott Cotton Mills. Richard Candee found paintings of industrial landscapes. Ed Layton shared his incomparable knowledge of turbine development. Joel Tarr added environmental insights, while Merritt Roe Smith helped me put this topic into a broader historical context. Hunter Dupree introduced me to the history of technology and systems theory when I was a graduate student and alerted me to the botanical pedigree of Kirk Boott's family.

Without skilled archivists like Martha Mayo and Joseph Kopycinski at the Center for Lowell History (formerly Special Collections) of the University of Massachusetts Lowell, Dan Walsh and Jack Herlihy at the Lowell National Historical Park, Laura Linard and Robert Lovett at the Historical Collections of Harvard Business School's Baker Library, Clare Sheridan and Helena Wright at the American Textile History Museum, and Peter Drummey at the Massachusetts Historical Society, we would not have the marvelous documentary collections that make in-depth research on Lowell's industrial history possible. I must also mention Janet Pohl and Janine Whitcomb, who always provided exceptional assistance at the Center for Lowell History.

There are no historical collections at the Tsongas Industrial History Center of the University of Massachusetts Lowell, but its working models and teacher workshops played an important role in the research for this book. Dorrie Bonner, Mike Folsom, Sheila Kirschbaum, Tim Lavallee, Peter O'Connell, Bev Perna, and other Tsongas staffers deserve great credit. While I was director at Slater Mill Historic Site, the museum board encouraged hydraulic investigations and let me contract with Walt Pulawski to assemble a full-size breast wheel (designed by Chuck Parrott) in the Wilkinson Mill. Running machinery with that wheel taught me more about the effects of backwater than all my documentary research.

Citations in the full set of online notes for this volume demonstrate my enormous debt to hundreds of writers and their publications, but I will single out for particular praise a few of the people who produced cultural resource reports, park plans, and exhibits that proved invaluable during this project: Jim Beauchesne, Anne Booth, Simeon Bruner, Pauline Carroll, Beth Frawley, Larry Gall, Carolyn Goldstein, Mark Herlihy, Susan Lyons, Betsy Bahr Peterson, Ellen Rosebrock, and Robert Weible. Whenever French technical literature was too difficult for me, I turned to Bruno Belhoste or Jeff Horn. For technical advice, I depended on many of the people already mentioned, as well as John Bowditch, Jed Buchwald, Pauline Desjardins, Greg Galer, Duncan Hay, Barrett Hazeltine, Kingston Heath, Don Jackson, Bill Johnson, Gary Kulik, Peter Molloy, Steve Mrozowski, Ted Penn, Michael Raber, Marty Reuss, Arnold Roos, Matt Roth, Giovi Roz, George Smith, and Ido Yavetz.

At the Johns Hopkins University Press, my editor, Robert J. Brugger, suggested and carefully guided this publishing project; Josh Tong and Kim Johnson got the manuscript into production; Wilma Rosenberger did the book and jacket design; and Claire McCabe Tamberino and Christina Cheakalos handled marketing copywriting and publicity, respectively. Barbara Lamb's meticulous copyediting whipped the final manuscript into shape. Two of my Brown

University students, Gregory Anderson and Rebekah Souder-Russo, helped me go over page proofs from the press. All of these people and many others are responsible for the positive things about the book. Only I am to blame for any errors.

My last thank you is to Fishing Creek, in Pennsylvania, where I wrote most of this book in rural isolation and learned painful lessons in hydrology. When the creek rose to unprecedented levels in 2006, I nearly lost my book manuscript, my family, and my life, but I also saw the power of water from an entirely new perspective.

Waterpower in Lowell

Introduction

Lowell, Massachusetts, was a city built on waterpower. Tourists were already flocking there by 1848, when a published handbook encouraged strangers to apply in advance for cards of introduction that would "insure a ready admission to the mills."[1] Ten well-financed corporations were using waterpower from a two-level canal system to make textiles for an expanding national market. Another firm on that five-mile network of canals turned out machinery. Watching the heavy waterwheels revolve in their stone pits was a highlight of any tour. It was also a thrill for visitors to stand in a weave room, absorbing the sounds and vibrations of densely packed, belt-driven looms. Mechanical power transmission linked the great wheels to the production machines, but the ultimate source of that energy, and the reason for Lowell's flourishing industry, was the Merrimack River.

Waterpower spurred the industrialization of the United States and was the dominant form of power for its manufacturing until well after the Civil War. Production centers grew at major drops on many American rivers, particularly those in the eastern half of the nation. These cities at the falls depended on power canals, specially constructed channels that delivered water to the prime movers (waterwheels or turbines) of industrial plants. When the first waterpower volume of the Tenth U.S. Census appeared in 1885, its general introduction made the following claim: "It is probably safe to say that in no other country in the world is an equal amount of water-power utilized, and that, not only in regard to the aggregate employed, but in regard also to the number and importance of its large improved powers, this country stands pre-eminent." The earliest and most successful of those "large improved powers" was at Lowell.

The location of Lowell at Pawtucket Falls on the Merrimack River was no accident. The tremendous waterpower potential of the site, only twenty-seven miles from Boston, attracted a massive surge of capital investment, beginning in 1821. Engineers, industrialists, and construction crews rapidly transformed a rural village, with clusters of small mills and a failing transportation canal, into a booming center for textile production and machine building. The canal system that they created, expanded, and continually improved was the principal supplier of power for Lowell's corporations until the mid 1880s. Canal routes had the highest priority in the design of the city and remained a prominent part of the urban landscape as Lowell matured. The city's economic prosperity drew a great deal of attention, inspiring both envy and emulation. Lowell was America's premier example of industrial development based on waterpower.

James Montgomery, a British textile expert who worked in and reported on the American cotton industry in 1840, said that "the principal manufacturing town in the United States is that of Lowell, which may justly be denominated the Manchester of America." He noted that "in general the Mills throughout the United States are moved by water; indeed, the water power resources of this country are incalculable, and many years must elapse before they can be fully brought into use."[2]

By 1848, Lowell's industrial plants were running 300,000 spindles, a good measure of productive capacity in textiles. The mills employed nine thousand females and four thousand males to produce a variety of products, including enough cotton cloth every day to "extend in a line about a yard wide" for two hundred miles. According to the 1848 handbook, Lowell had shown that "manufacturing on a large scale" could be profitable in America without "degradation" of the workforce. The promotional publication concluded that the city, "taken as whole, may be considered as a magnificent and successful experiment."[3]

Land speculators and aspiring industrial barons all over America dreamed of creating another Lowell. Waterpower visions often exceeded reality. John Quincy Adams predicted that Zanesville, Ohio, would become "the Lowell of the West." Henry Clay was even more optimistic, implying that waterpower from the Muskingum River made Zanesville the "best manufacturing site in the United States."[4]

Confidence in America's industrial capability had been much more subdued in the years immediately after the Revolution. At that time, few people believed that the new nation could compete with Britain in manufacturing. America's water-powered gristmills, sawmills, blast furnaces, and forges showed some promise, but mechanization of textile production lagged behind. The cotton "manufactory" that opened in Beverly, Massachusetts, in 1787 had serious limitations because it lacked a waterwheel. Beverly workers operated spinning jennies and looms with their own muscle power, while a team of horses walked outside on a circular track, turning the shaft for a carding machine.

The power demands of the factories that spurred American industrialization quickly exceeded the physical capabilities of humans or animals. Even wind power, which had worked well for gristmills, was not easily harnessed in sufficient strength and was too variable for the steady pace of factory work. America's first true factory, the 1793 Slater Mill in Pawtucket, Rhode Island, made cotton yarn with power provided by a waterwheel. Moses Brown, a Rhode Island merchant, financed the construction of that wood-framed mill, which was operated by Samuel Slater, an immigrant with experience in English textile manufacturing.

When Slater's in-laws, the Wilkinsons, erected a substantial stone mill next to his smaller wooden one in 1810, they installed a supplemental steam engine as well as a waterwheel. Slater later built a cotton mill powered entirely by steam on the Providence waterfront in 1827, but the high cost of fuel to generate steam made waterwheels the preferred prime movers for most New England textile mills until after the Civil War. Waterpower from rivers was an abundant natural resource in this region.

The most ambitious of the first waterpower schemes was situated further south, however, in Paterson, New Jersey. Alexander Hamilton and other investors in the Society for Establishing Useful Manufactures hoped to develop a large industrial community around the Great Falls of the Passaic River at what is now Paterson. Pierre L'Enfant, the French engineer who produced the influential city plan for Washington, D.C., created an urban design for the society in 1792 and envisioned a multi-tier canal system to power its manufacturing operations. With the Passaic River falling more than sixty feet in a short distance, it was possible to use the same water more than once, thus dividing up the substantial drop and making room for more mill sites. In time, water discharged from mills on an upper level canal would supply those on a middle level, whose wheel pits would empty into a third level. Unfortunately, the society built only one cotton mill before it had to give up its ambitious scheme to run its own manufacturing establishments. The struggling organization reinvented itself as a property developer and distributor of waterpower. With Peter Colt as superintendent, it had abandoned L'Enfant's street pattern and built a simpler canal system. Expanding that system and attracting enough independent manufacturers to make the costly investment profitable would not be easy. Paterson's economic growth was slow in the first three decades of the nineteenth century.

More rapid success greeted the Boston Manufacturing Company, which began making cotton fabric on the Charles River in Waltham, Massachusetts, in 1814. Francis Cabot Lowell and his fellow industrialists built a short canal to deliver water to a fully integrated textile mill. Their dam created a drop at the waterwheel of about eleven feet. For the first time anywhere, machines driven by waterpower took cotton through all the major stages of manufacturing, from raw fibers to woven cloth, within a single building. The Waltham power loom, which Lowell and his superintendent, Paul

Moody, based on British technology, was a key component of this factory system, in which material moved from process to process with clocklike regularity. By 1817, two brick mills stood in line between the power canal and the river, with a small machine shop nearby.

This unusual arrangement of mills and power canal was to become a model for many future efforts to harness large-scale water-power sites in America. Nathan Appleton, one of the founders of the Boston Manufacturing Company, had made a brief visit to a utopian mill village in New Lanark, Scotland, in 1810. He noted in his journal that he observed "a range of very extensive cotton mills through the windows of which we saw the wheels in full motion for 4 or 5 stories together."[5] The power canal paralleled the River Clyde below the falls, and two of the three mills then in operation were in a line between the canal and the river. F. C. Lowell was also in Scotland at this time, and although there is no evidence that he saw New Lanark, he must have discussed the place with Appleton. The layout in Waltham, which allowed more than one mill to draw water from the same canal and still benefit from the full drop, was too much like the New Lanark power system to have been accidental.

Most American textile mills in the early nineteenth century were perpendicular to the river or canal that powered them, a placement that worked well for a single building but could make it difficult to add more factories sharing the same drop. The parallel arrangement at Waltham suggests anticipation of future expansion. Indeed, it was relatively easy to extend the canal downstream to power the second mill in Waltham.

When waterpower and space proved inadequate to support a substantial expansion of operations in Waltham, the principal stockholders in the Boston Manufacturing Company looked elsewhere for a new manufacturing site. The place they chose in 1821 was then called East Chelmsford, but it became famous as the industrial city of Lowell.

This book tells the story of Lowell's creation and of the modifications to the canal system that were necessary to meet the increasing demand for power over time. The focus sometimes shifts to the larger Merrimack River basin, where industrialists from Lowell and other cities engineered changes in natural discharge patterns to provide more reliable flow in dry seasons. *Waterpower in Lowell* is a case study of urban industry and a history of technological choices. It credits the hard work and skills of the people who built and operated the Lowell Canal System and protected the vulnerable city from catastrophic floods, but it also reveals social and environmental costs associated with dependence on the river.

A central figure in this history is James B. Francis, the chief engineer and managing executive of the Proprietors of Locks and Canals. He may have been the most talented and capable engineer in America during the heyday of direct-drive waterpower. After 1845, his canal company was controlled by the manufacturers that it supplied with water: ten textile corporations and a large machine shop. He was a widely admired leader not only in the city of Lowell but also in the emerging profession of engineering. In the development of American scientific engineering, his role was crucial. *Lowell Hydraulic Experiments,* which he published in 1855, remains a classic in the technical literature. He was an engineer who distrusted pure theory, a careful investigator who preferred to base his conclusions on the observable results of large-scale experiments. He wanted his meticulous research to have practical value as a guide for other engineers. This erudite but largely self-educated man, for whom the "Francis" turbine is named, was also responsible for much of the corporate landscaping that contributed significantly to the image of Lowell as a model industrial community.

The six chapters that follow explain how the Proprietors of Locks and Canals harnessed the Merrimack River to power the mills of Lowell. Five of them cover specific historical periods up to 1885, and one is devoted to the subject of scientific engineering, including water measurement and turbine testing. The two-level canal system that Francis operated was unusually large and very complex. Controlling it was a difficult challenge. Technological innovations, many of them originating in the city, made possible better allocation and more efficient use of waterpower as the century progressed. At the same time, improvements in steam boilers and engines, combined with cheaper transportation of coal, reduced the cost advantages that waterpower enjoyed. Steam, when used first for power and then for processing or heat, was a bargain for some mills. Increasingly efficient steam engines and boilers also added reliability and extra energy to hybrid steam/waterpower systems, which were common in Lowell mills by the 1870s.

The detailed coverage in this book ends at 1885, but a postscript takes the story of waterpower in Lowell through the electrification of mills, the collapse of textile manufacturing in the North, and up to the present. The year 1885 can be considered a turning point in the city's history for many reasons. James Francis retired at the end of 1884, marking the close of an illustrious fifty-year career at Locks & Canals. Waterpower was still very important, but installed steam horsepower had surpassed it on the canal system. Improvements to the canals were relatively minor after 1885. Direct-drive waterpower had been essential for America's industrial development and a source of great national pride. Well-informed engineers like Francis recognized that its era was ending. Hydroelectric generation and the transmission of energy through wires were already possible in 1885. By the turn of the century, manufacturers who wished to use the power of falling water would no longer have to place their factories near a falls or depend on extensive canal systems like the one at Lowell.

For students of industrial history and for anyone who contemplates a visit to the Lowell National Historical Park, this volume provides a readily accessible introduction to nineteenth-century technology and to a place where America's manufacturing prowess was on impressive display.

CHAPTER ONE

Harnessing the Merrimack River

The Merrimack River flows down from the highlands of New Hampshire, past the great industrial cities of Manchester, Lowell, and Lawrence, and into the sea at Newburyport, on the northern coast of Massachusetts. The natural drainage basin of the river extends northward to the flanks of Mount Washington. Near that mountain's summit are the headwaters of the Pemigewasset River. Dozens of streams and lakes feed that river as it runs south to Franklin, New Hampshire. There the Winnepesaukee River joins it from the northeast, bringing the full discharge of Lake Winnepesaukee and its southern bays. The junction of these two rivers forms the Merrimack, 269 vertical feet above the mean tide at Newburyport.

In its one-hundred-ten-mile journey to the sea, the Merrimack does not give up its altitude in a smooth, uniform descent, but instead rushes forcefully over a number of falls, rapids, and manmade dams. One of the greatest changes in the level of the river occurs at Pawtucket Falls, just south of the New Hampshire line in Lowell. For centuries, these falls produced a natural descent of more than thirty feet. Early resident Henry Miles said that the drop was "not perpendicular, but over several rapids, in circuitous channels, with a violent current amidst sharp-pointed rocks."[1] A masonry dam now stands at the head of the falls, diverting much of the water into power canals but also increasing the total drop though the rugged stone formations that stretch for more than a half-mile downstream.

The river makes a turn to the northeast before entering Lowell, then bends sharply to the southeast after passing the rapids of Pawtucket Falls. It continues in that direction for a mile before the Concord River joins it from the south. Then, the Merrimack swings again to the northeast and the sea, some thirty-seven miles away. The falls, the bend, and the confluence of the two rivers in Lowell have greatly affected the history of the area.

The Pawtucket Falls have concentrated human activity for thousands of years. Native Americans congregated here because of the abundance of salmon, shad, alewives, eels, and sturgeon. When the Indians traveled to or from the interior on the Merrimack River they carried their canoes and cargo around the worst parts of the falls. The early English colonists also struggled with the laborious portage, although some risked life, limb, and property by running the rapids on downstream trips.

Making full use of the major falls on the Merrimack or the Concord was too ambitious for the early settlers of East Chelmsford, but a number of country mills used falling or rushing water to power their operations before the end of the eighteenth century. In 1691, John Barret had a clothier's mill for processing woolen fabric on River Meadow Brook. Another mill lot, laid out beside the brook in 1696, was the site of a sawmill by 1714. A detailed "Plan of Chelmsford," drawn in 1794, shows the clothier's mill with a sawmill and a gristmill nearby. The same map reveals a growing number of small-scale industrial ventures. An ironworks with a trip hammer appears at Massic Falls on the Concord, and two sawmills occupy sites closer to that river's junction with the Merrimack. On the Merrimack itself, there is a sawmill at the head of Pawtucket Falls and another one, in association with a gristmill, at its foot. One sawyer had an earlier mill beside the falls, but the great flood of 1785 swept it away.

For many of the eighteenth-century inhabitants of the Merri-

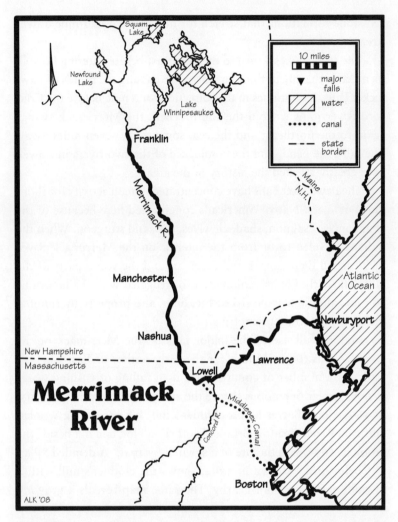

Map showing Lowell, the Merrimack River, the Middlesex Canal, and the principal New Hampshire lakes that served as reservoirs. Amy Kendall, delineator.

mack Valley, the steep drops on the main river looked more like an economic obstacle than an opportunity. Loggers, farmers, and inland manufacturers were particularly distressed by the liabilities of the river as an artery of commerce. Boats were easily smashed and their cargo scattered. Rafts of logs might float over some falls or rapids in high water, but the frightening cataract in East Chelmsford was an exceptional obstruction. Engineer James B. Francis wrote that the usual practice was "to break up the rafts at the head of Pawtucket Falls, haul them by ox teams to the foot of the falls, a distance of about a half a mile, and there re-raft them."[2]

Pawtucket Falls remained an intimidating threat for river travelers until a company called the Proprietors of the Locks and Canals on Merrimack River bypassed it with a canal. Assuming that there would be heavy traffic in forest and agricultural products from New Hampshire, the Proprietors began planning their Pawtucket Canal in 1792. Demand for masts, lumber, and naval stores was particularly high at that time in the busy port and shipbuilding center of Newburyport, at the mouth of the Merrimack, where almost all the investors made their homes. Hoping to make large profits by charging tolls for passage through the new canal, they commenced excavation the following year. The path they chose kept excavation to a minimum and capitalized on an existing pond and stream to help create a navigable waterway. The engineering problems were still difficult, however, because the canal had to move boats and large rafts through a thirty-foot change in elevation. The finished canal bypassed Pawtucket Falls, substituting smooth passage through a series of gravity-operated locks for the drop through rapids and jagged rocks. From the end of the artificial waterway, on the Concord, it was only a few hundred yards to the junction with the lower Merrimack.

Canals with locks to move vessels between different levels were new to America, but a number were under construction in the 1790s. The typical lock was a watertight chamber with paired miter gates at either end. Gates of this type (meeting at an angle pointing upstream) could be opened easily to allow boats to enter or leave the chamber from either direction. Also essential were valves (usually set in the miter gates) that could admit water to the closed chamber from upstream or discharge it downstream. The operation of a lock was simple. If the upper gates were closed and the lower gates were open, a boat going upstream could enter the lock. The tender of the lock then closed the gates behind the boat and opened the upstream valves to admit water into the chamber from the upper level of the canal. Gravity alone was all that was needed to fill the chamber until it matched the level of the water upstream. A boat rose quickly with the water in the chamber to the higher level. Once the water level was the same on both sides of the upstream miter gates, the lock tender opened the gates by hand and let the boat out. Operation for traffic heading downstream was similar, with vessels entering from the upstream level while the lock chamber was full of water. Discharge valves would then drain water from the closed chamber until the downstream gates could be opened for a boat to leave on the lower level. It was also possible to link locks in series, with boats going directly from one chamber to another for sequential changes in elevation (up or down a "flight" of locks).

Although only one and three-quarter miles long, the Pawtucket Canal deserves to be counted among the transportation landmarks in the early history of the United States. Hydraulic engineering on this side of the Atlantic at the end of the eighteenth century did not approach the best practices in England, France, Italy, or the Netherlands; American projects even drew the open scorn of some foreign observers. A French duke was surprised to see that the Great Dismal Swamp Canal was being constructed in 1796 without any levels being taken. He claimed that the builders began it without knowing "what number of locks may be necessary, and even whether any will be required."[3] Things were not much better on the Pawtucket Canal, where a surveyor laid out the route, but no professional engineer was employed.

The opening in October 1796 was not exactly a triumph. The directors' minutes refer only to a "disagreeable incident." A more dramatic account was provided by the Reverend Wilkes Allen, who wrote that "the occasion had called together a great concourse from the vicinity." As the crowd was watching a boat pass through the first lock, the sides of the structure "suddenly gave way" and "water bursting upon the spectators with great violence, carried many down the stream." Although some had their clothes "almost entirely torn off from them," no one was seriously injured in this mishap.[4]

This collapse took place at what became known as the "Swamp Locks" (then a single lock chamber). Reconstruction of the failed lock and delays caused by winter weather held off the commercial use of the canal until the next spring. The lower set of locks (originally two chambers) may also have been rebuilt before full operation began. Total costs of construction for the canal amounted to approximately fifty thousand dollars. Costs had risen so high by 1797 that the Proprietors successfully petitioned the state legislature for higher rates of toll than had been authorized in 1793. They argued that the canal had cost more than twice what they had planned to spend.

More than two locks (or sets of locks) were necessary to make the canal a safe and effective transportation corridor. The Proprietors quickly built the "Guard Locks" to protect the waterway from rising water in the river and then added the "Minx Locks" to help with passage over some high ledge downstream of the Guard Locks. Both of these structures had only one lock chamber. There were also changes at the "Lower Locks," which became a flight of three chambers. All these modifications were completed by 1802, but major repairs and improvements continued until at least 1810. An 1821 map by J. G. Hales shows the canal in considerable detail.

The venture paid dividends averaging about three dollars per share per year from 1797 to 1807, despite problems with shallow sections, water shortages, a series of damaging floods, and increasing competition. In its first decade of operation, it was not an economic disaster, but better investments were available. After the Embargo of 1807 and the War of 1812 had put New England's overseas trade into a depression, there was less demand for timber and other forest products in Newburyport.

By 1821, the Pawtucket Canal was not in good shape, either structurally or financially. Its wooden locks were deteriorating and much of its business had been captured by the Middlesex Canal, which opened in 1803, or the Middlesex Turnpike, which opened

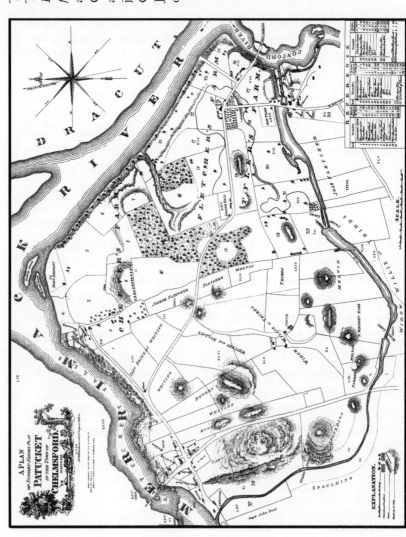

The place that became Lowell. This *Plan of Sundry Farms Etc. at Patucket in the Town of Chelmsford, 1821* shows Pawtucket Falls at the far left. The Pawtucket Canal runs in an arc from just above the falls to the downstream junction of the Merrimack and Concord rivers at the far right. J. G. Hales, delineator. Courtesy of the Lowell Historical Society.

in 1810. The Middlesex Canal provided another way for water-borne cargo to reach the sea. It was designed by Colonel Laommi Baldwin with the help of English engineer William Weston, who used an imported spirit level (a framed rotating telescope with a rigidly attached bubble level to adjust for perfectly horizontal sightings) during the surveys.

The new canal left the Merrimack a mile upstream of Paw-tucket Falls and offered a direct, twenty-seven-mile passage to Boston Harbor. Twenty locks and eight aqueducts answered the topographical challenges of the route. Partly as a result of this competing waterway and partly because of the decline in trade and shipbuilding in Newburyport, shipments down the lower Merrimack had diminished markedly. With difficult rapids still remaining on that stretch of the river, there had never been any significant upstream traffic from the coast. The immediate prospects for the aging Paw-tucket Canal were not promising.

Twenty miles away, in Waltham, the much shorter canal of the Boston Manufacturing Company was a different type, with only one purpose: to supply waterpower to the mills of that firm. Close proximity to the port of Boston was an asset, as was the high level of technological skill and innovative spirit among artisans and managers at the site. One of the interesting features of the Waltham production system was its automatic control of waterwheel speeds. Mechanical regulation through the use of a centrifugal governor first appeared in eighteenth-century windmills and was then applied to the throttle regulation of early steam engines. Similar feedback devices soon proved effective in the water-powered mills of Britain. The spinning flyballs of these governors were driven by the main shaft in a mill and, through variations in centrifugal force, could "sense" changes in machine speeds. Increased speed drove the heavy balls further apart. If workers turned off some of the textile machines, the waterwheel and linked shafting would start to speed up because the power would exceed the load. The governor would immediately react by beginning to close the gate that provided flow for the water-wheel. Once proper machine speeds had been restored, the governor went back into neutral position, leaving the gate in its new setting. The same process worked in reverse to increase the flow of water onto the wheel whenever more power was needed. Workers could turn on and off machinery as required during the workday and know that their governed waterwheel would quickly adjust to the changing power demand in the mill.

F. C. Lowell had observed waterwheel governors during his visit to Britain before he opened his first textile mill in Waltham. According to the nineteenth-century textile historian William Bagnall, Lowell told Moody, who was his chief machine builder, that "they must have a governor, to regulate the speed of the wheels." After giving Moody a very limited description of this device's form and operating principles, he ordered the skilled mechanic to get one from Britain. Later, when Lowell checked to see whether Moody had made the purchase, he got a negative answer. He was surprised to learn that Moody had instead made one in the machine shop. Bagnall noted that the governor ran until 1832 and "was the model of those afterwards used in Lowell."[5]

Moody's action demonstrates a common method of transatlantic technological diffusion that deserves more attention from historians. Alert observers on pleasure tours, purchasing trips, or travels to cultivate business markets often saw machinery and industrial processes that were unknown in America. Upon their return, they described what they had seen to artisans, like Moody, who had the technical ability to convert their mental images into physical reality. In some cases (such as F. C. Lowell's borrowing of English power loom technology), this was deliberate industrial espionage, but not always. The transmitter of new knowledge might not intend to steal the innovations of others. Once word of a technological advance had reached receptive ears, someone was likely to put it to use. Ideas with significant promise could stimulate simultaneous or sequential efforts by more than one artisan. Another great machinist, David Wilkinson, was making and selling waterwheel governors in Rhode Island by the early 1820s. It is possible that gifted men like Wilkinson and Moody may have added their own incremental innovations to this feedback mechanism, thereby improving a device that was not their own "invention." Waterwheel governors became more sophisticated by mid century.

Less successful than the flyball governor was an American invention that F. C. Lowell hoped would eliminate the effects of backwater on the high breast wheel in his Waltham mill. Backwater occurred whenever streams, swelled by rain or snowmelt, rose to flood the wheel pits of mills along their banks. Waterwheels lost efficiency or slowed to a halt when they had to turn through high water. Breast wheels had become popular with New England manufacturers in the late eighteenth century because they handled moderate levels of backwater better than overshot wheels. Their direction of rotation, opposite that of overshot wheels, went with the flow of water out of the wheel pit, rather than against it. They could push some water away, but too much backwater would still hinder a wheel's performance. This problem forced many mill owners to set their breast wheels a foot or more above normal tailrace levels, giving up drop (and therefore power) they could not afford to lose. In 1814, as he was completing his new mill, Lowell chose to install a "machine for removing the backwater."[6] It was designed by Jacob Perkins, who had once employed Moody.

Perkins's machine operated something like the nozzle of a fire hose. It took advantage of the extra flow available during floods by directing a jet of water to the point where the breast wheel discharged into the pit. The kinetic energy of this high-velocity blast was supposed to blow water away from the wheel and keep the tailrace from refilling the wheel pit. Although Lowell liked the initial results enough to buy half the patent rights to Perkins's invention, he could not persuade many people of its worth. Most waterpowered mills continued to suffer from backwater, and Perkins's machine joined the long list of clever ideas that did not work well in practice.

Despite the slight setback with the backwater machine, the experiment in water-powered textile manufacturing and machine building at Waltham was a clear success by 1821. The Boston Man-

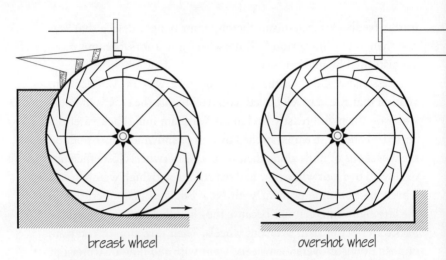

breast wheel   overshot wheel

Advantage of the breast wheel over the overshot wheel in backwater. The direction of rotation of the breast wheel helped push backwater out of flooded wheel pits. Backwater was a greater impediment to the operation of the overshot wheel. Amy Kendall, delineator. Based on figure 5-8 in Terry Reynolds, *Stronger Than a Hundred Men* (Baltimore, 1983).

ufacturing Company's corporate organization, high level of capitalization, innovative labor system, and focus on the mass production of plain cotton cloth had carried it through the economic depression that followed the end of the War of 1812 and the return of British competition. Measured by return on investment, it was doing much better than the Society for Useful Manufactures at Paterson, New Jersey. With a new protective tariff in place and demand growing for inexpensive fabric, F. C. Lowell's faith in mass production seemed justified. His death in 1817 had not affected the momentum of the enterprise, but all the readily available waterpower at the Waltham site was already in use, and a planned expansion into printed calicoes would require a new site with much more power.

Industrial enterprises in East Chelmsford in 1821 were far more modest than the thriving operations in Waltham. The small agricultural community, with about fifty houses and a population of at least two hundred, surrounded by swift flowing rivers and a decaying transportation canal, had been making steps toward a diversified economic base that included more water-powered manufacturing. A local sawyer, grist miller, and fuller named Moses Hale added wool carding to his repertoire and then branched out into gunpowder production. In 1820, he left the explosives business in the hands of his partner, Oliver Whipple. After asking engineer Laommi Baldwin II (Harvard-educated son of the builder of the Middlesex Canal) to confirm the feasibility of his plan in 1821, Whipple dug a power canal parallel to a steep section of the Concord River. This half-mile canal, like the shorter one in Waltham, was more ambitious than a traditional raceway, or "trench"; it soon supplied power not only to Whipple's enlarged powder works but also to a series of smaller mills, some of which he rented to independent entrepreneurs.

Manufacturing also clustered around the Lower Locks of the Pawtucket Canal (see page 14). Joseph Tyler, the first contractor for the transportation project, made an arrangement to use water from the canal to power a sawmill in the 1790s. When the War of 1812 cut off English competition, thus creating a "hot-house environment" for textile producers, John Goulding and Jonathan Knowles started a mill nearby on the Concord, where they spun cotton, carded wool, and made narrow fabric webbing. In 1818, they sold the property to Thomas Hurd, who made woolen cloth there, and he added a three-story brick mill with a long headrace from the river in 1821. Goulding may have supplied most of Hurd's machinery, including power looms, from a machine shop that he set up at the Lower Locks in 1815. Using canal water leased from the Proprietors of Locks and Canals, Goulding operated machine tools there until 1821. After leaving the area, he went on to become a famous inventor of woolen machinery, including the "Goulding Condenser."

As winter approached in 1821, East Chelmsford was no longer just a rural community with acres of rocky farmland and a few country mills. It was a place with possibilities, already affected by the forces of industrialization and unsure of its future. Some outside investors with ties to the busy textile factories in Waltham were already considering the prospect of harnessing the entire power of Pawtucket Falls on the Merrimack River. Before the decade of the 1820s was over, this village at the falls would experience one of the most dramatic transformations in urban history.

CHAPTER TWO

Building a City at the Falls, 1821–1836

If you were looking for a site to build an industrial city near Boston or Newburyport in the fall of 1821, Pawtucket Falls in East Chelmsford had to be high on your list of possible locations. Yet Nathan Appleton, in his 1858 memoir, *Introduction of the Power Loom, and Origin of Lowell*, gives the impression that he and Patrick Tracy Jackson had difficulty finding an available site with adequate waterpower for a large manufacturing enterprise. Their first scouting expedition, to southern New Hampshire, turned up nothing appropriate:

> Soon after our return, I was at Waltham one day, when I was informed that Mr. Moody had lately been at Salisbury, when Mr. Ezra Worthen, his former partner, said to him, "I hear Messrs. Jackson and Appleton are looking out for water power. Why don't they buy up the Pawtucket Canal? That would give them the whole power of the Merrimack, with a fall of over thirty feet." On the strength of this, Mr. Moody had returned to Waltham by that route, and was satisfied of the extent of the power which might be thus obtained, and that Mr. Jackson was making inquiries on the subject.[1]

According to Appleton, Jackson corresponded with the agent of the "Pawtucket Canal Company" and "ascertained that the stock of that Company, and the lands necessary for using the water power, could be purchased at a reasonable rate."[2] The two entrepreneurs could hardly have been unaware of the potential of Pawtucket Falls or the financial condition of the Proprietors of Locks and Canals, the corporation that owned the decaying transportation canal around the sharp drop in the Merrimack. Jonathan Jackson, the first president of the canal company, was, after all, the father of Patrick Tracy Jackson. The son had inherited much of his father's canal stock. The mercantile community around Massachusetts Bay certainly knew of both the Pawtucket and Middlesex Canals and of the great falls they bypassed.

The idea that the falls were little known and their potential unappreciated persisted. Noted orator Edward Everett, speaking near the site on July 4, 1830, praised the "natural capital" represented by American waterfalls and blamed British colonial policies for delaying their widespread utilization in the eighteenth century. He implied that no one had recognized the industrial possibilities in East Chelmsford: "The site of Lowell itself was examined, no very long time before the commencement of the first factories here, and the report brought back was, that it presented no available water power."[3]

Whatever the reputation of Pawtucket Falls in the fall of 1821, by the time the first snow flurries were swirling over the river, a growing number of ambitious men realized that East Chelmsford had a great deal of "available water power." John A. Lowell said that Jackson envisioned a "much more stupendous project—nothing less than to possess himself of the whole power of the Merrimack River at that place."[4] Since Jackson was too busy running the mills in Waltham, and Appleton was apparently not interested in taking on the vast responsibilities of creating a power system and setting up a manufacturing company, they needed to find a suitable chief executive for the new venture. Jackson said that Kirk Boott, a former British Army officer and member of a prominent Boston mercantile family, would be interested in such a position and "had the proper talent for it." The eager young man quickly agreed to join them.[5]

Once Jackson, Appleton, and Boott had decided to develop the site in East Chelmsford, they moved expeditiously to take control of the Proprietors of Locks and Canals and to acquire land before word spread about their ambitious plans. Riparian property with water rights at the falls was the highest priority, but they also had to have enough land to build a large number of mills and house thousands of workers. Thomas Clark, the agent of the Proprietors of Locks and Canals, agreed to make purchases for them, doing so as quickly and quietly as possible to avoid a rise in prices. His good relationships with local landowners helped move the project forward. He took "the deeds in his own name, in order to prevent the project taking wind prematurely," but Boott kept all the accounts.[6]

With land and stock transactions under way, the men behind the grand plan traveled to the site. Describing their inspection on a cold day in late November, Appleton wrote that "we perambulated the grounds, and scanned the capabilities of the place, and the remark was made that some of us might live to see the place contain twenty thousand inhabitants."[7]

Efforts to take control of the canal produced rapid results. Stock was already changing hands by November 14, 1821, when the directors of the transportation company officially heard what developers had in mind. In the directors' minutes, Boott is identified as the agent of a company planning "to establish a large mill works in a manufactory." His letter, read at that meeting, "expressed a desire to purchase of the Proprietors . . . all the mill power they own at Chelmsford."[8] Boott soon followed that up with several cash proposals, but the takeover was really accomplished by stock purchases, with Clark again playing a major and initially secretive role. The developers had enough shares to give Boott and his associates four of the five directorships of the company on December 26. Boott soon held key administrative positions as both agent and treasurer of the Proprietors of Locks and Canals. His principal power base was, however, in the newly formed textile-manufacturing corporation known as the Merrimack Manufacturing Company.

Boott had been offered the position as agent (managing executive or chief operating officer) even before the company was formally organized on December 1, 1821. He and his brother John together controlled as many shares of the corporation's stock as any of the five subscribers, who included Appleton, Jackson, and Moody. Within a week, nine other individual stockholders were added by transferring some of the original six hundred shares. Shares also went to the Boston Manufacturing Company, many of whose proprietors became active in the East Chelmsford development. The total amount of stock represented a very large supply of available capital ($600,000), as assessments up to $1,000 per share were allowed in the articles of agreement. At the first stockholders meeting of the Merrimack Manufacturing Company, on February 27, 1822, Boott became clerk and treasurer, as well as agent. The treasurer was the most powerful officer in these early textile corporations.

The new company was generally successful in acquiring East Chelmsford property inexpensively. Before rumors of a significant industrial development had spread, it held title to most of the land between the Pawtucket Canal and the Merrimack River, as well as sections adjacent to and south of the canal. One individual, however, created difficulties. Thomas Hurd, who, as we have seen, had a woolen mill on the Concord River, learned of Boott's plans in time to buy up land and water rights that the Merrimack Manufacturing Company needed. On November 27, 1821, he purchased the Bowers sawmill at Pawtucket Falls, just before Boott was about to acquire it. He later bought other parcels of land in or beside the falls, and he drove a hard bargain with Boott for all of these important properties. Boott's company had to pay him $18,000, a very high price at the time, for less than twenty-eight acres and some mill privileges.

The industrialists who wanted to build a great manufacturing center now controlled more than four hundred acres of land in East Chelmsford, the old Pawtucket Canal, and all the potential waterpower from Pawtucket Falls. From the beginning, they were thinking about a large number of textile mills, but siting the first of these new factory buildings was a key decision, one that would affect the layout of power canals and, indeed, of a whole community. Of the early investors, Paul Moody had the most experience with water-powered manufacturing, and he had earned the confidence of Appleton and Jackson with his technical achievements as superintendent at Waltham. Appleton later explained that "it was decided to place the mills of the Merrimack Company where they would use the whole fall of thirty feet. Mr. Moody said he had a fancy for large wheels."[9]

The topography of East Chelmsford and the shape of the old Pawtucket Canal placed limitations on the industrial and urban planning of the Merrimack Manufacturing Company. The ideal way to supply a number of mills with waterpower was to use a single canal running parallel to a river with a falls. This was the concept proven effective at New Lanark and at Waltham. If the canal left the river above the falls and re-entered at some distance downstream, then the land between the canal and the river became an extended island on which mills could be placed in a line. By keeping the level of the water in the canal close to that of the river above the falls, there was a major difference in height between the canal and the river at every point below the falls. Water from the canal could enter the mills on the island to drop through power-producing machinery, such as waterwheels, and then exit into the lower river. In this way, the potential energy of the water due to its elevation, or "head," could power manufacturing processes in each mill.

In East Chelmsford, neither the riverbed nor the adjacent land was suitable for such a simple solution. The drop in the river continued for almost a mile below the head of the falls. On the east bank, rocky terrain rose steeply from the river. The builders of the original Pawtucket Canal had avoided this high ground by running their channel in a wide arc around the bend in the Merrimack. If mill foundations were seated higher than the level of the river above the falls, water would not flow by force of gravity into the wheel pits. Nor could a new canal be cut parallel to the river without addressing the problems of elevation and solid ledge. Mills might line the bank of the Merrimack, but they would have to be down-

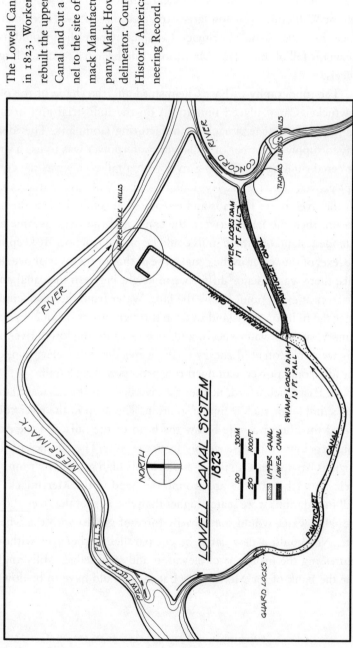

The Lowell Canal System in 1823. Workers have rebuilt the upper Pawtucket Canal and cut a new channel to the site of the Merrimack Manufacturing Company. Mark Howland, delineator. Courtesy of the Historic American Engineering Record.

stream of the rapids. There, past the bend, the full thirty-foot drop was available and the land surface was lower. Boott would have to select mill sites away from the falls, where canals could reach them.

To hold down construction costs and get textile production under way as quickly as possible, Boott and the other corporate directors decided to rebuild the Pawtucket Canal as a feeder for additional power canals. It would also remain as a transportation canal, both to satisfy the charter of the Proprietors of Locks and Canals and to carry cargo for the mills. Construction supplies, raw materials, and manufactured products would pass through the improved canal for many decades to come. In the spring of 1822, hundreds of laborers began to enlarge the upper Pawtucket and dig the new Merrimack Canal. The latter would run 2,600 feet, from the old Swamp Locks to a site chosen for the first of the Merrimack mills, on the bank of the river.

Boott personally supervised this challenging project. Moody, still superintendent at Waltham, helped with some of the early engineering but could not spend as much time on the site as could the resident agent. Jackson, who had gained technical experience while agent of the Waltham mills, also took an active interest in the canal work and mill construction at East Chelmsford. Boott's engineering management received close scrutiny during this initial phase of construction. Unafraid to take command, he soon earned a reputation as an imperious and demanding taskmaster.

There is no evidence that Boott had any engineering training before he took on this job. Some have claimed that he studied the subject at Sandhurst, the English military academy, but that was not the case. Boott's well-connected father had obtained, or purchased, a commission for his son in a light infantry regiment. His service, which included combat in the Peninsular War, did not involve construction. His later work as a civilian in Boston was in the family's merchant house.

Hydraulic engineering was still a young field in America in the 1820s. Few men were well trained in it, although large projects like the Erie Canal provided practical experience for beginning engineers. Soon after becoming agent of the Merrimack Manufacturing Company, Boott corresponded with one of the Erie Canal's well-known engineers, Myron Holley. In a letter written in December 1821, Holley gave him advice on the construction of dams and locks. Boott also looked at some of the literature on French hydraulic engineering, then the best in the world. An undated mailing to his Boston home provided him with information about du Buat's method for calculating velocity in rivers and canals.

Practical considerations seem to have been more important than theory in actually rebuilding the Pawtucket Canal. Boott and Moody apparently wanted a cross section of 480 square feet. They intended to create a uniform channel, sixty feet wide by eight feet deep, but crews working on the Pawtucket Canal were not able to cut or wall that precise rectangular shape. Most of the channel looked more like a trench than a precise work of engineering. There were vertical sides in some places and sloping ones in others. Surfaces included cut ledge, masonry walls, loose stones or gravel, and earth. Difficult rock formations were sometime left jutting into the canal because of the urgent pace of construction, and the width of the waterway exceeded sixty feet in many places. Boott was apparently willing to compromise in order to remain on schedule. Almost sixty years later James Francis said that the canal was still "irregular and has many bends."[10] Branch canals were also hard to make; tough quartzite caused particular problems for men who excavated the line of the Merrimack Canal.

Most of the five hundred or more laborers who worked on the reconstruction of the upper Pawtucket and the digging of the new canal to the Merrimack mill were hired by contractors who had made agreements with Boott to complete specified tasks. Nearby agricultural areas and small towns supplied not only large numbers of men but also teams of horses and oxen. Charles Blaney and Hugh Cummiskey brought gangs of Irish workers from the coast. Cummiskey left Charleston, Massachusetts, with a crew of thirty and, after walking almost twenty-seven miles to East Chelmsford, arrived too late and too tired to begin digging. According to the local tradition, Kirk Boott followed Cummiskey's suggestion and bought the men a drink in a nearby tavern. Refreshed, or perhaps numbed, they grabbed their shovels and went to work the next morning.

Dr. John Green, physician to the canal laborers, remembered that much of the excavation in the spring of 1822 was for the new branch canal, later named the Merrimack Canal, and for the foundations of the first mill. The workers had barracks near Dutton Street until early June, when the corporation moved the expanding labor force to two large barns "fitted up with bunks for sleeping, and rough tables for feeding."[11] These temporary facilities were the first company housing that the Boston Associates provided in East Chelmsford.

As the Irish came in increasing numbers, a distinctive Irish community began to form. The Merrimack Manufacturing Company soon lost interest in providing even rudimentary housing for day laborers. Most of the Irish workers took shelter in makeshift huts on unused pieces of land in what became known as "the paddy camps." Some housed their families in this uncomfortable and unhealthy setting. Dr. Green recalled waiting on the sick in "their wretched dwellings . . . with an umbrella over my head to shield me from the dripping of a leaky roof."[12] The paternalistic regard for textile operatives that made the "Lowell System" famous did not extend to the day laborers, who struggled to build the canals, the first mills, and the neat rows of corporate housing. Whether Yankees or Irish, they usually had to fend for themselves.

The geographic restrictions of water-powered industrial development often forced manufacturing companies to build permanent housing for their regular workforce. This was particularly true in Rhode Island, where most sites with adequate power for a textile mill were scattered through the rural hinterlands on streams of small to medium size. Mechanized production was only feasible where there was a natural fall or rapid with enough flow for year-round operation. Industrialists had to create mill villages in places where settlement had previously been sparse. Company ownership of housing gave managers much control over workers' lives, but it was an expensive investment with continuing maintenance costs. In rural East Chelmsford, as at the falls in Waltham, the need to attract large numbers of female operatives, as well as male managers and skilled mechanics, left the Boston Associates little choice but to provide good housing in an attractive urban setting.

Work on the Pawtucket and Merrimack canals was physically demanding and very dangerous. Some six thousand pounds of gunpowder were used to blast ledge. Men worked with hammers, hand drills, shovels, and wheelbarrows along the canal bottom while teams hauled out tons of earth and rock. Stonemasons laid walls where necessary but often relied on cut or natural ledge to form part of the canal's sides. In the swampy sections the canal had to be narrowed and strong banks formed to restrain the flow. This muddy work of filling and walling was as onerous for the workmen as excavation. Filling continued for years in the low areas further down the Pawtucket Canal.

Dr. Green, who treated many of the workers, realized that "the nature of their employment exposed them to much accidents and disease." His medical practice was "much among them, night and day," and he was the first to be notified when men were hurt. On July 8, a worker was "blown up by gunpowder in blasting rocks." Other "similar and frequent accidents" followed. Russell Mears had charged two holes and was tamping the powder in a third "when it exploded, throwing him into the air. He fell near the other holes, which exploded in succession, and no one could approach until afterwards." Mears lost both an arm and an eye "and was pierced in all directions by the fragments of stone, some of them weighing two ounces." Injuries were so common that Green carried out "five amputations of arms" in two years.[13]

Despite these industrial hazards, the construction crews made steady progress. Moody's plan for using the whole thirty-foot drop at the Merrimack Manufacturing Company meant that one intermediate lock (the Minx Lock) on the existing Pawtucket Canal had to be removed. Now there would be no significant change in water level from the river to the Swamp Locks, where the Merrimack Canal began. Boott had workers rebuild the Swamp Locks and raise the adjacent dam at that central location. His plan called for a drop of thirteen feet at the Swamp Locks, leaving seventeen feet of head for the lower level of the Pawtucket Canal.

Boott chose to build strong stone lock chambers, but he did not try to make their masonry walls leakproof. Instead he lined them with wooden planks, whose joints would swell tight to prevent water from escaping. The wood lining could be replaced as needed, but the stone walls were built to last. The Guard Locks had just one chamber, for only during high water was there any appreciable drop between the upper river and the upper level of the canal. At the Swamp Locks two chambers were required to move boats or rafts thirteen vertical feet. Similarly, the even larger drop to river level at the Lower Locks called for two chambers.

The Guard Locks had protected the old transportation canal from floods. With the upper Pawtucket serving as a feeder for power canals, the role of the Guard Locks became more complex. The miter gates of the old lock chamber had spanned the channel, blocking the flow of water from the river whenever they were closed. This was not a problem for a transportation canal in which there was little current. A power canal, on the other hand, cannot have even brief blockage of its flow without suffering a rapid drop in water level. If a canal was to serve as both a transportation and a power system, the operation of the locks could not interfere with the constant flow of water to the mills.

Boott solved this potential problem by creating a separate channel around the rebuilt guard lock. To protect the canal from floodwaters and to regulate flow, he placed sluice gates across this bypass. Employees could adjust the gate openings to satisfy changing demands for water. With this modification at the guard lock site, Boott gave priority to the delivery of energy for manufacturing and began the creation of a controlled power system that could handle much greater flow than the old towpath canal.

Boott and Moody had the first Merrimack mill nearing completion just as the water began flowing to it on September 2, 1823. Two days later Boott wrote a significant passage in his diary: "Thursday, September 4, 1823. After breakfast, went to factory, and found the great wheel moving round his course, majestically and with comparative silence. Moody declared that it was 'the best wheel in the world.' N. Appleton became quite enthusiastic."[14]

The wheel was similar to the original wheels at Waltham but on a much larger scale. It was a high breast wheel, thirty feet in diameter with buckets in two sections, each twelve feet wide. This type of wheel operated with the force of gravity: the weight of water in the buckets produced power. Constructed of both wood and iron, the "great wheel" was controlled with adjustable gates automated by a flyball governor. As in other mills of this period, a combination of gearing and heavy iron shafting transmitted mechanical power from the revolving wheel to each room in the mill. This is one variation of what is sometimes called "direct drive," to differentiate it from the later development of electrical generation, wiring circuitry, and motor drive.

Some of Moody's artisans from Waltham had been on the scene in East Chelmsford for months, helping to build the new mill and install its machinery. Standing out among these experienced craftsmen was John Dummer, born in 1791 and already a noted millwright. At the Boston Manufacturing Company, he was "foreman of the water-wheel and pattern work, etc." Moody's son, David, knew him well: "In the year 1822, John Dummer and myself went up the Merrimack River, near Nashua, brought down a raft of logs, to the Stony Brook saw-mill, and there got out the lumber for the first two wheels of the Merrimack Mill." William Worthen, an engineer who worked in Lowell in the 1830s, said of Dummer, "As a millwright, he was the best I ever knew. . . . He built the first wheels at Lowell in 1822, and none of them were, I think, ever renewed."[15]

Before the first mill was built or the Merrimack Canal completed, Nathan Appleton had compiled an estimate of the industrial potential of the Merrimack Manufacturing Company. In this document, he listed the progress anticipated by March 1, 1824: "7,200

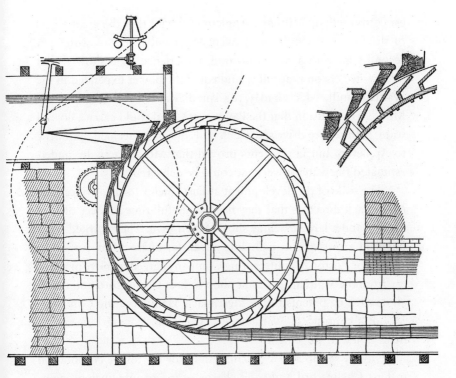

A section view of a Lowell breast wheel and its governor. The flyball governor at the upper left controls three sliding gates, which admit water to the buckets of the wheel. A close-fitting breast, or apron, keeps water from leaving the buckets until the bottom point of rotation. Reprinted from Joseph P. Frizell, *Water-Power* (New York, 1905), pl. 127.

spindles in 2 factories of 3,600 each: a bleachery and printing establishment."[16] These structures were part of a much grander plan, but the company didn't quite meet this rapid pace of construction.

Appleton had conferred with Moody before making predictions about waterpower and manufacturing capability. Moody estimated that canals with thirty feet of head could "furnish power to carry 60 factories of 3,600 spindles each." Appleton, "to be on the safe side," planned for fifty such mills (with "all the apparatus for weaving, etc.") in addition to the two already in construction.[17]

These men were dreaming of up to sixty factories like the largest cotton mill at Waltham. Appleton thought that the great wheel of the first mill would need twenty-four cubic feet of water per second (cfs) with a thirty-foot fall. Experience soon showed that twenty-five cfs on that fall would equal the power expended in the Waltham mill, which had 3,584 spindles and a full set of looms. Moody's conclusion that the Pawtucket Canal could carry a flow of water sufficient to drive sixty such mills may have seemed optimistic to Appleton, but later events proved that Moody actually underestimated the waterpower that could be delivered.

The scale of the development at East Chelmsford convinced the Merrimack directors that they needed the full-time services of Paul Moody and a large machine shop of their own. They worked out a financial arrangement (including a transfer of stock) by August 9, 1823. The Boston Manufacturing Company agreed to move machinery from the Waltham machine shop to East Chelmsford and to sell all its existing patents to the Merrimack Manufacturing Company. Moody's contract in Waltham was canceled, leaving him free to make a new arrangement.

Moody set up a new machine shop for the textile corporation in East Chelmsford and took charge of its manufacturing operations. William Worthen remembered that the machinery "was made after the designs of Paul Moody . . . and no change was allowed to be made by any one except with his approval."[18] The shop was near the Swamp Locks, on a triangle of land between the upper-level Merrimack Canal and the lower level of the Pawtucket. By digging raceways from the Merrimack Canal to wheel pits in the shop, then discharging the water into the lower Pawtucket, Moody used the thirteen-foot difference in water levels between the two canals to generate mechanical power.

Choosing to use two levels of canals opened up a much larger area for water-powered manufacturing; all mills would not have to be on the banks of the Merrimack. Mills on the upper level (other than those of the Merrimack Manufacturing Company) would discharge water into a lower-level canal, where it would be used again before going back to the river. The same water would pass through

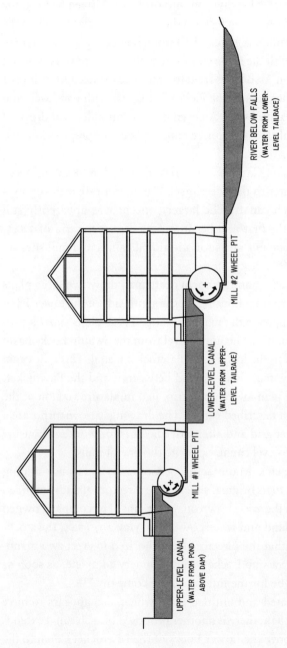

Concept for a two-level canal system. This schematic diagram depicts an upper-level canal carrying water from the pond above a dam to a mill. Water used for power in that building discharges through a tailrace to a lower-level canal, which then delivers it to another mill. The two drops in the diagram match those established in Lowell. After its second use for power, the water returns to the river below the dam and falls. Charles Parrott, AIA, delineator.

mills on each level. This division into drops of thirteen feet (on the upper level) and seventeen feet (on the lower level) was based on the existing topography. A hatched contour line on Hale's 1821 map identifies land with the same elevation as the river above Pawtucket Falls. The Boston Associates tried to minimize the need for filling or excavation as they rebuilt the Pawtucket Canal with two levels and selected potential mill sites. An even division into identical drops of fifteen feet would have required much more site preparation and work on canals.

The Pawtucket Canal was not reconstructed in 1822 and 1823 just to bring water to the Merrimack Canal, but rather to supply a system of branch canals. The larger concept was apparently still evolving when the first mill began production at the end of 1823, although someone put ideas for additional canals and mill sites on paper in 1824.

In January 1824, an unknown draftsman drew two large plans for mills and corporate housing on the south side of the lower Pawtucket Canal, opposite the machine shop. These plans show a proposed canal running at the upper level from the Swamp Locks basin in a line parallel to the lower-level Pawtucket Canal. This new canal created an elongated island of land between it and the Pawtucket. The mills in the plan are on this strip, in a miniature version of the ideal power plan described earlier. The drawings also incorporate a rectangular street grid and alternative layouts for workers' housing across the upper-level canal, opposite the line of mills.

The Merrimack Manufacturing Company was busy adding additional mills to its original site, but the directors already intended to build more in the area shown on these plans. The company owned a great deal of land and waterpower. On May 19, 1824, the stockholders voted "that the directors be authorized to erect two manufacturies at the Swamp Locks with machinery complete, as soon as they judge it to be for the interest of the company."[19]

On an undated and untitled plan, which also appears to have been drawn in 1824, there is another proposed branch canal extending from just above the Lower Locks of the Pawtucket Canal to the area near the junction of the Concord and the Merrimack. It is a lower-level waterway designed to reach mill sites along the Merrimack River southwest of the Merrimack Manufacturing Company. It could also supply mills along the western bank of the Concord downstream of the Lower Locks. This future canal, later to be called the Eastern, would provide a seventeen-foot drop.

Despite all the planning for new canals and mills, by the fall of 1824, the directors (who were also the principal stockholders) of the Merrimack Manufacturing Company were having second thoughts about managing these greatly expanded waterpower and manufacturing operations. It was hard enough to run a textile corporation without the added problems of a machine shop, a dam, transportation locks, multiple mills, and an expanding network of power canals. No one anywhere had experience harnessing such a great waterpower. This dwarfed New Lanark, the largest British site. The directors may also have perceived a risk to their initial financial success if they continued to tie new enterprises directly to one company. If one or more of the planned ventures should fail, the losses would hurt the corporation as a whole.

In October, the directors set up a committee to see whether it was "expedient to organize the Canal Company" and to sell it the canals, water rights, land, and buildings that the Merrimack Manufacturing Company did not need.[20] The 1792 charter of the Proprietors of Locks and Canals was still valid, and the older corporation could take on additional responsibilities, including the continuing development of the industrial community at East Chelmsford.

While the committee investigated new roles for the Proprietors of Locks and Canals, the Merrimack directors considered a plan to let another textile company join them in East Chelmsford. On November 22, 1824, they decided to make a bargain with the Hamilton Manufacturing Company, a newly formed textile business, "for the sale of one or more mill powers, including machinery and land at the rate of thirty dollars per spindle, and for the sale of any mill powers with land, without machinery, for four dollars per spindle."[21]

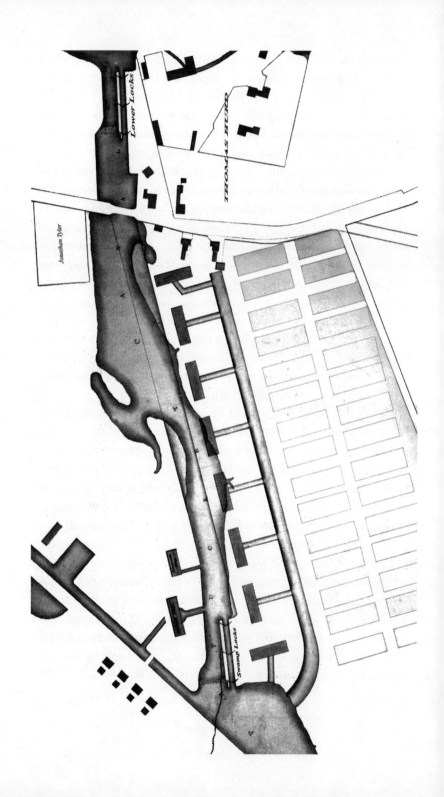

At the same time, they began discussions with William Lawrence, another potential purchaser.

The term *millpower* (or *mill power*) referred to the amount of water power required to drive a textile mill, in this case the second mill at Waltham. That factory had 3,584 spindles and all the other machinery necessary to turn raw cotton to woven cloth. As we have seen above, each of the two Merrimack mills had machinery approximately equal in power demand to the Waltham standard, and the Hamilton Manufacturing Company also planned to build mills of similar capacity. Long after the relationship between a millpower and the actual number of spindles it could drive lost any real meaning, leases were still by spindle at the Waltham ratio. Although later mills were bigger and needed more than one millpower, the term remained in use.

For the Merrimack directors, a millpower was 25 cfs of water on a thirty-foot drop. That was equivalent to 85 gross, or theoretical, horsepower. With an estimate of 64 percent efficiency for the high breast wheels in use, the net, or usable, power was 54.4 horsepower. The Hamilton Manufacturing Company leased a flow of water with only a thirteen-foot drop, from the upper-level Hamilton canal to the lower level of the Pawtucket Canal. An increased flow of 60.5 cfs per millpower compensated for the lower drop and the slightly reduced efficiency of wheels with a smaller diameter. On the final drop of seventeen feet, the flow in later leases would be 45.5 cfs. The millpowers at each level were supposed to represent the same amount of energy for manufacturing. Many other communi-

(*Opposite*) A plan for new mills along the lower Pawtucket Canal in 1824. This is an ideal plan, with a new upper-level canal parallel to the existing lower-level canal and a line of water-powered mills between them. Blocks for the boardinghouses of mill workers are across the upper-level canal in a rational grid. This plan also shows the intention to straighten the Pawtucket Canal downstream of the Swamp Locks. Image enhanced for this publication. Courtesy of the Center for Lowell History, University of Massachusetts Lowell.

WATERPOWER

To calculate waterpower, we multiply the weight of flowing water per second times the head, or drop, at the wheel. We will use a weight of 62.4 pounds for each cubic foot of water in our example.

The Proprietors of Locks and Canals leased water in units called *millpowers*. Each millpower was equivalent to the power used to run the second mill at Waltham, which had 3,584 spindles and all the other machinery necessary to turn cotton into cloth. One millpower at the Merrimack Manufacturing Company, which had the canal system's full drop of 30 feet, was determined to be 25 cubic feet per second (cfs).

Using those figures, we can easily calculate how much gross (theoretical or potential) horsepower each millpower represented.

Waterpower =
flow rate in cfs × weight of water in pounds per cubic foot × head in feet

One millpower =
25 cfs × 62.4 pounds per cubic foot × 30 feet =
46,800 foot pounds per second

One horsepower is 550 foot pounds per second, so we divide our answer by 550.

One millpower = 46,800 ÷ 550 = 85 gross horsepower

In reality, no waterwheel could capture all the gross horsepower from the water. Efficiency was the ratio of the actual horsepower captured to the gross horsepower. A breast wheel like the ones at the Merrimack Manufacturing Company had a normal efficiency of about 64% (0.64). If we multiply that efficiency times the gross horsepower, we can estimate the net (actual) power available for manufacturing:

85 gross horsepower × 0.64 efficiency = 54.4 net horsepower (for a Lowell breast wheel). Turbines, introduced in Lowell in the 1840s, had higher efficiencies. A measured efficiency of at least 80% (0.80) was not unusual with turbines. Applying that efficiency to our 85 gross horsepower (one millpower), we get 85 gross horsepower × 0.80 efficiency = 68 net horsepower (for a turbine).

ties adopted the term *millpower,* but there were sometimes local variations in its meaning. Millpowers from Pawtucket Falls were always based on the specific flow rates and drops established in the mid 1820s. They were leases of water, not power. The actual amount of shaft horsepower that a mill developed from one millpower would increase over time for reasons that will be explained in later chapters. In addition to the initial purchase of water rights and land, each corporation had to pay an annual "rent" of $300 per mill-power. In this way, the Proprietors of Locks and Canals were treating water as a commodity.

The location of the proposed Hamilton mill complex was beside the lower Pawtucket Canal, on part of the land set aside for factories in the 1824 plans of the Merrimack Manufacturing Company. Prominent investors in the older textile company helped form the new one. Patrick Tracy Jackson explained that the Hamilton mills would not compete with the Merrimack mills because their product would be "twilled cotton goods—a different article from anything we have yet attempted." He also predicted that "all our proprietors, who feel able, will take stock in it."[22] Thus the familial, social, and mercantile connections that had linked these investors since their first manufacturing venture in Waltham continued to influence the patterns of business expansion. Historians give the name *Boston Associates* to men from the capital city who put up money and provided managerial direction for Lowell corporations and later for similar ventures in other New England communities.

In 1825, the Merrimack directors arranged to transfer the machine shop, the remaining undeveloped land, and the waterpower rights to the Proprietors of Locks and Canals. Now there would be one company that would sell land and lease waterpower to new textile-manufacturing firms. The Merrimack Manufacturing Company would own just one of many mill complexes on the canal system. Of course, stock subscriptions gave the owners of the original textile corporation a chance to share in the profits of the revitalized canal company (commonly known as Locks & Canals).

As part of the arrangement, the Merrimack Manufacturing Company sold its riparian property on both sides of Pawtucket Falls to Locks & Canals. The intention was for the canal company to control all of the water coming down the Merrimack River. It would then lease water to a select group of paying customers on its power delivery system. In its "indentures," or contracts, with the manufacturing corporations, Locks & Canals agreed to supply a specified number of millpowers during the working day and to build and maintain the canals necessary to do that. This was considered nothing more than a rental of water. The arrangement conveyed no riparian water rights in the traditional sense. The textile producers did not acquire a specified fraction of the river's flow, as would mills that bought property beside a waterfall on a natural stream.

Manufacturing corporations had to pay for their own head-races to bring water from the canals to their wheels, and for their own tailraces to discharge that water into a river or lower-level canal. Since most new companies paid Locks & Canals to erect and equip their mills, they usually chose to let the canal company build their head gates, trash racks, races, wheel pits, waterwheels, and power transmission systems as well. A steady progression of building projects and lucrative machinery orders made Locks & Canals a highly profitable company.

Sale of land to new corporations also generated a great deal of income. This was surely one of the motivations for planning a canal system on two levels. Dividing the fall created more sites with potential for water-powered manufacturing and thus increased the value of real estate in East Chelmsford. Each textile corporation would need significant space for its mills, storehouses, repair shops, and worker housing. The Hamilton Manufacturing Company, following the example of the Merrimack, also wanted land for its own fabric print works.

A highly detailed map from 1825 shows the land holdings and buildings of the Merrimack Manufacturing Company just before the official transfer of property to Locks & Canals. It also lays out for the first time the proposed routes for an entire system of waterways on two levels. This remarkable document, delineated by consulting engineer George Baldwin, includes every power canal that would be constructed by 1836. Baldwin had learned engineering and drafting on his own, with help from his well-known father

(Laommi) and brother (Laommi II). His 1825 map suggests a comprehensive solution to the problem of power distribution. It incorporates the principal features of the 1824 plans described earlier and adds more canals. Here is evidence that the system was carefully planned at an early date and was not the result of multiple decisions made over a fifteen-year period. There was nothing haphazard about the placement of canals in this industrial community.

Kirk Boott was supposed to make the 1825 plan a reality. Despite what would today be considered a conflict of interest, he remained agent and treasurer of both the Merrimack Manufacturing Company and Locks & Canals. Paul Moody, now residing in Lowell, became the chief engineer of the latter. Moody took responsibility for the machine shop, the canals, and the engineering of various construction projects. Boott managed the sale of property and waterpower and was deeply involved in any business disputes concerning water rights.

One controversy at the falls was particularly frustrating for Boott because it presented a threat to his plans for a permanent dam at Pawtucket Falls. Boott was sure that he had purchased all the existing mill privileges for the Merrimack Manufacturing Company, but a challenge occurred late in 1824 and was still unsettled when the textile company transferred its water rights and land holdings to Locks & Canals. Boott had been using the former Bowers sawmill's wing dam on the east side and possibly a second wing dam on the west side to keep the first Merrimack mill in full production. Then another wing dam of unknown origin appeared below the main falls.

To Boott's extreme dismay, the unexpected dam construction on the west side turned out to be another enterprise of the redoubtable Thomas Hurd, the same man who had bought up land and mill privileges in 1821 and had forced Boott to buy them at inflated prices. This time Hurd had acquired an overlooked privilege from a local farmer, and he was threatening to sue if Boott cut off any of his water with a new dam at the head of the falls.

Negotiations grew heated in the winter of 1826. When Hurd

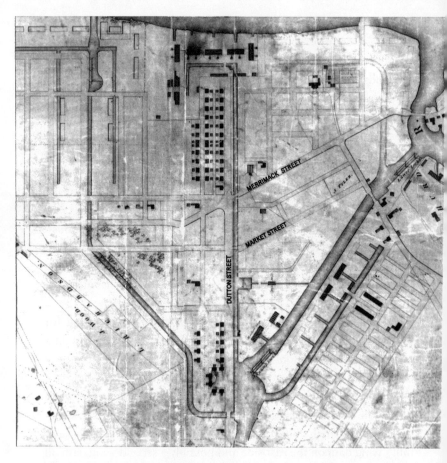

Detail from a plan for buildings, streets, and a system of canals on two levels, drawn by engineer George Baldwin for the Merrimack Manufacturing Company in 1825. It would take until 1836 to complete the canals that Baldwin envisioned. Street labels have been added for this publication. Courtesy of the Lowell National Historical Park.

started to build a mill on the river, Boott got corporate approval to "remove part of the whole of the dam" that would supply that mill.[23] Luckily, that drastic and probably illegal action was not necessary to get rid of the problem. The directors of Locks & Canals finally agreed to give Hurd a cash settlement, a land exchange, and the right to buy two millpowers at $12,000 each. Once again Hurd came out on top in a dispute with Boott, and he was probably not disappointed to see the agent's men demolish his work in progress below the falls. One of Hurd's new millpowers was to be drawn from the canal supplying water to the Hamilton Manufacturing Company. The canal company allowed Hurd to build a flume from the end of the Hamilton Canal to his own canal along the Concord River. A few years later he purchased his second millpower, which he drew directly from the lower Pawtucket Canal just above the final locks. He thus became the only manufacturer to get power from both the upper and lower levels of the canal system.

Hurd was not the only one to cause troubles for Boott at the falls. In 1825, the Merrimack Manufacturing Company built a temporary dam extending all the way across the river, for the first time blocking the main channel in East Chelmsford. The new structure probably incorporated most of the older Bowers wing dam. Witnesses said that the temporary dam was built mainly of stone slabs supported by timber. In some places, workmen had probably used timber framing or cribwork to close natural chutes between the jagged rocks. This temporary dam at the falls must have infuriated local fishermen because it threatened the annual runs of salmon, shad, alewives, and sturgeon in the river.

On September 20, 1825, Boott wrote a newspaper advertisement in which he offered a $100 reward for information leading to the conviction of the persons who did "maliciously tear up a part of the dam in Merrimack River, at the head of Pawtucket Falls." He was still advertising in 1827. A half century later, Daniel Varnum and several other residents still remembered the incident. They said that fishermen had simply pulled off some slabs from the temporary structure, "the shad, being seen above the dam, trying to find a passage for their return to the sea."[24]

Blocking the natural movement of anadromous fish to the sea has received less historical attention than obstruction of their upstream spawning runs. In some cases the trip downstream could be intimidating or impossible. This was particularly true with shad, the most important food fish of eastern rivers in the nineteenth century. Schools of shad are known to be easily confused and are likely to swim in circles when faced with an unusual obstacle, such as Boott's dam.

Fish normally went up rivers to spawn in the spring or early summer, when high water provided numerous passages at most falls and reduced the heights that they might have to jump. Powerful salmon can leap over dams as much as eleven feet high. Low dams, which are often completely submerged in freshets, presented little challenge to these athletic and highly determined fish. Where dams did represent a serious obstacle, mill owners and ironmasters sometimes provided a rudimentary bypass or opened sections of their dams temporarily to let the fish through. Ample water supplies from snowmelt and spring runoff allowed them to do that without significant disruption of manufacturing.

In the late summer and fall, water levels were usually much lower, but surviving adults and juvenile fish heading to the sea could still get over natural falls by finding chutes or overflow points. An industrial dam was different; it could halt their progress and trap them in its millpond. Mills regularly drew so much water during the workday that the level of the millpond would fall below the top of the dam. If flow was minimal, the pond might not fill up overnight or even on Sunday. Nothing would spill over the dam for days or weeks in some cases. Mill owners or waterpower companies would try to conserve every drop. All the flow coming down the river would be held in the pond or diverted into a canal or headrace, which was often some distance upstream of the dam. Fish approaching a dam that blocked the natural channel would find no way over or around it. This was apparently what happened in Lowell in September 1825, when fishermen intervened to create some spillage over the dam for the shad. It seems unlikely that the damage to the dam was as terrible as Boott's ad implied, but the direct action against company property was an injury he could not ignore.

In England, the common law remedy of direct action against a dam that flooded upstream property or blocked fish might have been permissible, but not in the state of Massachusetts, where a series of "mill acts" in the eighteenth century had made unauthorized dam breaking illegal. American law became increasingly supportive of water-powered industrial development in the national period. Although owners of some dams still had to open passages for fish in certain seasons or pay damages for economic losses they caused to established fishermen, the threat of vigilante action was diminishing.

The legal climate that encouraged water-powered industrialization had negative effects on New England's native fish. Sturgeon were prominent among the species that suffered as dams spanned the Merrimack. Scholars believed that the river was in fact named for the majestic, slow-moving fish, which looked like a survivor from the age of dinosaurs. They thought that *Merrimack* was one spelling of the Algonquian word *monomack,* which meant sturgeon. It was no coincidence that the weather vane on the first Merrimack mill was in the form of a sturgeon. It is ironic that the corporation named for the Merrimack River, and for the sturgeon that swam in it, was the first in Massachusetts to put a dam across that great artery and thus injure its fisheries.

Significant changes in both the physical and political landscape were soon under way in 1826. Work on a permanent dam commenced that summer, right after the flourishing industrial community of East Chelmsford became the incorporated town of Lowell. Named in honor of the man who had formed the Boston Manufacturing Company and built its dam in Waltham, Lowell was busy creating the hydraulic infrastructure that would make it the envy of the nation.

The reorganized Locks & Canals now had full responsibility for dam construction. Since there was no longer any need to channel water to the former site of the Bowers sawmill, the permanent dam could take a shorter line from the eastern shore to the "Great Rock." Some of the western sections of Boott's temporary dam may have served as a cofferdam, protecting the new construction downstream. The permanent dam relied on large timbers, gravel, and stone masonry to provide mass for resisting forces exerted by the river. Masons set the heavy stonework in place without mortar. The Great Rock, which projected above the crest, became an integral part of the dam, tying it firmly to solid ledge and giving the structure a dramatic outline when it was completed in 1830. Hoping to avoid further conflict with fishermen, Locks & Canals included a "fishway" with this dam. The rough bottom of the channel meant that some parts of the timber cribwork extended to greater depths than others, but in general, the dam was not very high. Pawtucket Falls and its downstream rapids still provided most of the thirty-foot drop for the canal system.

Three years later, the canal company increased the height of the dam another two feet, with two courses of granite headers applied to its crest. A remarkable oil painting shows this effort in progress. Workers also raised the fishway at the same time. The dam was now at thirty-two feet, according to the Locks & Canals scale of heights. The millpond rose with the dam, creating backwater problems for the Jackson Manufacturing Company at the junction of the Nashua and Merrimack rivers. Although prominent Boston Associates were the principal investors in that textile company, they got little sympathy from the directors of Locks & Canals or the corporations it served. Lowell needed more power, and if an upstream mill had to suffer, that was unfortunate but unavoidable. Eventually a cash settlement eased some of the bad feelings.

A much improved dam, more available waterpower, and more canals (either completed or in construction) supported the industrial expansion that continued to draw people to Lowell. Already there had been a phenomenal increase in population: from an estimated 200 in 1821, to 2,500 in 1826, to 6,477 in 1830 and to 12,963 in 1833. Reflecting the labor policies of the textile mills, there were far more women than men in Lowell. Neat rows of boardinghouses for female operatives stood close to the mills and canals.

The mills and most of the newer boarding houses were made of brick. A prosperous brick-making industry grew up in southern New Hampshire to fill contracts in Lowell. After observing the clay

A painting by Alvan Fisher of work on the Pawtucket Dam in 1833. The granite headers added at that time raised the millpond, which stretched for eighteen miles upstream. Jagged rocks give some sense of the geology at Pawtucket Falls. Courtesy of Harvard Business School.

pits and kilns on a river voyage in 1839, Henry David Thoreau commented that "from Bedford and Merrimac have been boated the bricks of which Lowell is made."[25] Vessels heavily laden with thousands of bricks could enter the canal system from the river and float directly to construction sites.

The highest priority of the investors in Lowell was to structure an environment conducive to manufacturing. Despite their business-like approach to the planning of a new city, these men also gave serious attention to the landscaping of mill yards, canal banks, house lots, and streets. Most of the company green spaces were open to the public, and all were visual assets for the city. The founders apparently believed that an attractive and tasteful urban setting reflected well on them and served to deflect criticism of American industrialism. It also helped recruit a proper workforce for their mills and boosted employee morale. Corporate landscaping efforts in Lowell were not examples of industry dominating nature but were instead the planned integration of natural features and deliberate plantings in the rational, highly structured order of a new industrial city.

Many American cities that contained power or transportation canals treated them simply as utilitarian infrastructure. Often they were hidden behind industrial and commercial buildings or fenced off, with little or no public access. The sensory appeal of these waterways was sometimes lost because urban wastes fouled their flow or because owners failed to maintain canal-side property. In nineteenth-century Lowell, the situation was different. Several of the constructed waterways served as showpieces for the community. Both corporate officials and the general public took pride in the aesthetics of the canal system, an urban network that would earn Lowell its reputation as the "American Venice."[26]

The Merrimack Canal was much more than a functional conduit for water. It was also an early and influential experiment in the beautification of American cities through landscape improvement. Corporate employees laid out Dutton Street parallel to the canal and put boardinghouses on the west side of the street. Kirk Boott, as agent of both the Merrimack Manufacturing Company and the canal company, directed the first landscaping of this corridor, an effort that extended into the Merrimack mill yard. With his tasteful plantings, he established a pattern of green engineering by Lowell corporations that would last long after his death in 1837. It is not surprising that Boott liked to see trees, grass, shrubs, and flowers on company land: horticultural and botanical interests ran deep in his family.

Kirk's father (Kirk Sr.), the son of an English market gardener, maintained a famous greenhouse and garden in Boston. One brother, Francis, became a respected botanical expert on sedges. All four of Kirk's brothers were members of the Boston Society of Natural History, and two, Wright and William, were charter members of the Massachusetts Horticultural Society at its founding in 1829.

The intensive planting of trees along the Merrimack Canal corridor apparently began north of Merrimack Street, on the strip of land between Dutton Street and the water. The new elms stood directly in front of the boardinghouses of the Merrimack Manufacturing Company. An account from 1825 described the setting: "On the banks of the factory canal which is fenced in and ornamented with a row of elms, are situated the houses for the people."[27]

While people strolled beside the Lowell canals, boats and rafts moved cargo through them, including a great deal of timber, brick, firewood, and coal. The Pawtucket Canal, serving as a transportation channel, carried things essential for the growing city. The canal's primary function now was to supply waterpower, but income from tolls was welcome, and local manufacturers appreciated the relatively low cost of waterborne transport.

Many of the textile mills shipped products out and brought raw materials in by canal boat. Competition from teamsters, who could use turnpikes to improve their trips between Lowell and Boston (and who could operate in the winter when canals froze), drove down the rates of toll on the Middlesex Canal. The canal's superintendent saw "the necessity of giving every encouragement to Manufactures."[28] Still, the textile industry seemed to be the savior of that waterway, which had paid no dividends to its investors from its 1803 opening until 1819. The locks in the power canals at Lowell made possible direct water connection between the Middlesex Canal and the mills. By 1832, the Middlesex was doing relatively well, and its prospects were looking better.

Because Locks & Canals kept leasing millpowers to new or expanding corporations, the flow into the system steadily increased, and with it the current in the Pawtucket Canal. As we will see, this caused serious problems in power delivery, but it also affected transportation. Travel against the current became more difficult over time, and boatmen found it easier to get into the city than to get back out. Even with the use of a reconstructed towpath, draft ani-

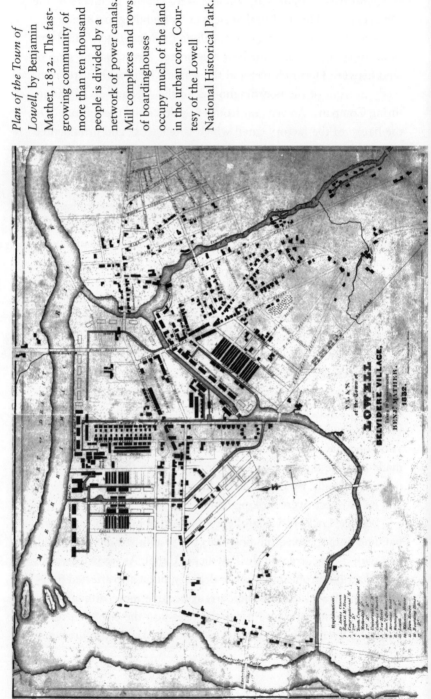

Plan of the Town of Lowell, by Benjamin Mather, 1832. The fast-growing community of more than ten thousand people is divided by a network of power canals. Mill complexes and rows of boardinghouses occupy much of the land in the urban core. Courtesy of the Lowell National Historical Park.

mals had to work hard to return boats to the entrance of the Paw-tucket Canal. Most of the arriving rafts, on the other hand, were broken up in the canal and their log components delivered to steam-powered sawmills conveniently located on waterside lots. Locks & Canals did its own sawing with waterpower at the machine shop. Some timber, masts, and naval stores continued through the city and down the Merrimack to Newburyport.

Expansion of the power canal system had to proceed quickly because the Boston Associates were eagerly incorporating new tex-tile companies. The Lowell Manufacturing Company, which came to focus on the production of carpets, built its mills north of the Locks & Canals machine shop in 1828. Its source of water was the new Lowell Canal, really little more than an extended headrace off the Merrimack Canal. Filling was necessary in this swampy area to create a level site with the proper elevation. The corporation's breast wheels operated with thirteen feet of head and discharged into the lower-level Pawtucket.

At the same time, Boott began another long canal that had two levels, like the Pawtucket. Opened in 1831, the Western Canal was soon providing power for three corporations. First the flow was split evenly to supply Tremont Mills and the Suffolk Manufacturing Company, on either side of the upper level. Tailraces from these cor-porations then emptied into the lower-level Lawrence Canal, which led to the Lawrence Manufacturing Company. Thus, the water had already been used once for power generation on a thirteen-foot drop before it reached the seventeen-foot wheels of the Lawrence mills.

After the Appleton Company opened on a site fed by the Ham-ilton Canal in 1828, the lower level of the Pawtucket Canal began to resemble an industrial canyon lined with factories and punctu-ated by tailraces. Some of the earth and rock from the Western Canal had helped to fill the coves and wetlands on the north side of the canal. This land-making effort included one of the earliest American uses of animal-drawn cars on a rail line. The idea of re-ducing friction with rails may have come from the 1826 Granite Railway at Quincy, Massachusetts.

The Lowell Canal System in 1836. The canals envisioned by George Baldwin in 1825 have been completed, and the Boott Cotton Mills are under construction. Mark Howland, delineator. Courtesy of the Historic American Engineering Record.

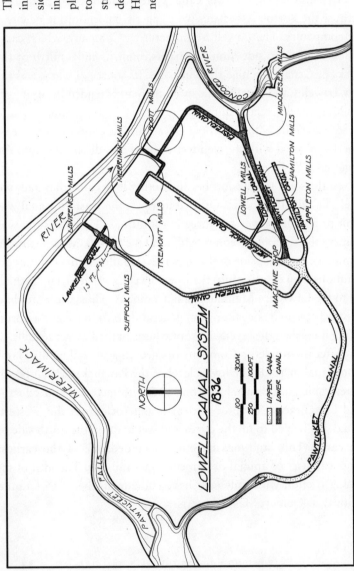

LOWELL CANAL SYSTEM
1836

Now the Locks & Canals machine shop and three textile corporations (Hamilton, Appleton, and Lowell) were discharging water into the Pawtucket from the upper level, but the only downstream purchaser of power was the Middlesex Manufacturing Company, a maker of woolen goods that had taken over Hurd's mills along the Concord by 1830. Locks & Canals needed more customers for its water on the lower level. For more than a decade, energy regularly went to waste over the dam at the Lower Locks.

Wasted power was a concern inside the mills as well. Every mill superintendent wanted to reduce energy losses in the direct-drive systems. The close connections between Locks & Canals and the textile firms helped them work together to increase efficiency in power transmission. Manufacturing corporations naturally sought to get as much production as possible out of each millpower they leased. Employees of the canal company shared their practical experience in hydraulic and mechanical engineering with the users of their water and the purchasers of their machines.

John Dummer, the millwright of Locks & Canals, was continuing to earn high praise for the reliable wheels he installed in most of the Lowell mills. He was a traditional artisan rather than a trained engineer, but he knew how to construct high breast wheels and, with Paul Moody, was probably most responsible for the sophisticated gate mechanisms and governors that kept them running at prescribed speeds. At first the prime movers, governors, gearing, and shafting departed very little from accepted British practice. Then Lowell began to benefit significantly from incremental innovation in its workplaces and the adoption of new technology. Secrecy remained an important concern in textile printing and dyeing operations but seems to have been less of an issue in power generation and transmission.

Ithamar A. Beard, superintendent of the Hamilton Manufacturing Company, ran a series of tests on the power used to drive machinery in his No. 2 mill in 1830. He published his observations three years later in the *Journal of the Franklin Institute*, the best-known technical publication in America. For these experiments,

Dummer constructed a special gauge wheel in the tailrace and put a closely fitted concave apron under it. Water discharged from the mill's three linked breast wheels filled spaces of known volume between the radial floats of the gauge wheel. After counting revolutions during a specified period of time, Beard calculated the flow rate and the gross power it represented at that site. He concluded that it took 96.34 cubic feet of water per second on a thirteen-foot drop, or 78,276 foot-pounds per second of power, to operate the entire mill. A dynamometer designed and operated by Lowell resident and mathematician Warren Colburn then allowed him to compare that gross figure with the mechanical power actually generated. Beard found that the combined breast wheel efficiency ("ratio of effect to power expended") was 0.6049.[29] In other words, almost 40 percent of the potential energy in the water was lost in the normal operation of the waterwheels.

At that time, the directors of Locks & Canals were not as interested in testing wheel efficiencies as they were in learning how much water was actually flowing to the mills. Beard's 1830 test of a single mill suggested that all the textile corporations might be getting more than they were paying for. Using findings from that measurement and extrapolating to assess the needs of an entire complex, Kirk Boott persuaded the Hamilton Manufacturing Company to lease two additional millpowers from Locks & Canals in 1832. The canal company made similar arrangements with most of the other mills that were already in operation.

Some of the Hamilton's gearing and shafting was already obsolete by the time Beard ran his tests. In setting up one of the Appleton Company's mills in 1828, Moody abandoned the heavy bevel gears and vertical shafts that had delivered power to each floor in earlier Lowell factories. Mechanical friction losses were high with these iron components, and it was difficult to fabricate or replace them. Damage to one of the gears or main shafts could shut a mill down for days or weeks. Moody installed a single leather belt of great length which transmitted power to horizontal line shafts on upper floors from a main drum driven by breast wheels in the basement.

This was not a perfect solution, but it was an important step toward the use of multiple belts running to drums or pulleys on the line shafts of particular floors. Smaller belts running from these shafts powered counter shafts and machines. Multiple-belt drive systems were relatively easy to repair and were widely adopted in American textile mills before the Civil War. Additional developments such as loose-running belts, sticky dressings, and higher shaft speeds further reduced energy losses over time.

We have one letter that shows how quickly another corporation in Lowell moved to a power transmission system with multiple belts. Aspiring mill superintendent Robert Israel gave a detailed description of wheels, gears, drums, and belts in the Merrimack mills in 1831, which by then had "no upright shafts." He praised the water-wheels in Lowell, which were "the most perfect I have yet met with." The breast wheel for a single Merrimack mill was really "two water wheels with a twelve foot drum between them." Gears cast in segments and bolted to the rim of each wheel drove the drum, from which belts ran to production floors. Israel said that "three main belts carry the whole mill" and that this means of power transmission was "much less expensive and upon the whole the best way."[30]

A Lowell textile mill functioned as a carefully regulated production *system*. Andrew Ure provided a classic definition of the factory system in Britain: "the combined operation of many orders of work people . . . tending with assiduous skill a system of productive machinery continuously impelled by a central power."[31] Operatives guided materials through sequential stages of production, with one set of machines feeding the next. A substantial level of automatic control was built into the power train and the machinery, but supervision by professional managers was still essential. The Boston Associates had readily applied British concepts in their own system.

As we have already seen, the "central powers" in Lowell's first textile mills were breast wheels. Mechanical drive systems, made up of a combination of gears, shafts, pulleys, and belts, transmitted power from these prime movers to the textile machinery. The rate of water flowing into the buckets of the waterwheels set the pace of

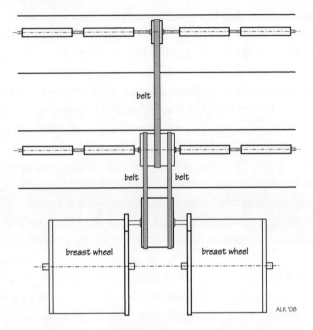

Multiple-belt drive in a textile mill in the 1830s. This is the mechanical power transmission system developed in Lowell. Gearing on breast wheels (or turbines, in a later period) drove the main pulley in the basement. Large leather belts ran to line shafts on upper floors. Smaller belts (not shown here) went from the line shafts to machinery on the same floor and through the ceiling to machinery on the floor above. Amy Kendall, delineator. Based on an illustration in Ithamar Beard, "Remarks on the Mode of Gearing Mills for the Manufacture of Cotton and Woolen Goods," *Journal of the Franklin Institute* 19 (1837): 455.

production. Flyball governors automatically adjusted this flow, with minimal lag, to keep machinery running at or near a predetermined speed. Mill operatives contested the occasional "speed-ups," which increased both productivity and job-related stress. Workers were much more than "cogs in the machinery," but their labor was clearly linked to the wheels.

Boott and other corporate agents depended heavily on mechanically adept men like Moody, who had tested the first American governor at Waltham. His technical aptitude and artisanal skills helped create the canal system and improve the use of waterpower in Lowell's textile mills. Several years before his death in 1831, Moody acquired a young assistant named Joel Lewis. The new employee had studied mathematics and taught in a school in Lowell. Lewis soon became the principal surveyor of Locks & Canals. He worked frequently on hydraulic engineering projects before a fatal illness curtailed his contributions in November 1834.

Another young Lowellian with mathematical talents stopped working for Locks & Canals in 1834 and began to concentrate on a career as an independent engineer. Uriah Boyden had already produced plans of a number of mill yards, surveyed much of the city, and mapped the river between Lowell and Nashua. He would go on to become one of the century's greatest inventors of hydraulic equipment.

Boyden left the canal company just as a young British immigrant was joining it. James Bicheno Francis was born on May 18, 1815, in Southleigh, Oxfordshire, England. In 1828, he moved to Wales, where his father had been appointed superintendent of the Porth Cawl Railroad and the harbor at its terminus. In a brief autobiographical statement, Francis noted that his father's work "naturally led the son to adopt engineering as a profession." When he was only fourteen, he left school to work on the construction of the impressive harbor works at Porth Cawl. Two years later, in 1831, he helped to build sections of the Great Western Canal in Devonshire and Somersetshire, under the direction of chief engineer James Green. By the time he immigrated to America at the age of seventeen, Francis already had a great deal of practical experience in engineering and construction. He said later that 1833 had been a "favorable time" to arrive because "the railway system was just then fairly starting." After working briefly for Phelps, Dodge & Company (later one of America's largest copper producers) in New York, he "found employment as an assistant engineer under William Gibbs McNeil and George W. Whistler, on the New York, Providence, and Boston Railroad."[32]

Whistler, a graduate of West Point, probably encouraged Francis to study mathematics and learn French, the language in which many of the best technical books of the period were written. There could not have been many better places to further one's education and practice in engineering than the relatively short stretch of railroad between Providence, Rhode Island, and Stonington, Connecticut, in 1833 and 1834. Whistler and McNeill became legendary figures in the founding of the American engineering profession, but they were not the only talented technical people around Francis at that time. On the same field crew building that early railway were the young American Julius Adams and the recent Scottish immigrant James Kirkwood. Francis, Adams, and Kirkwood would go on to be presidents of the American Society of Civil Engineers.

In the spring of 1834, Whistler accepted the post of engineer for Locks & Canals. He was probably planning to concentrate on locomotive building. Although he was also in charge of the waterpower system, he left much of the canal work to Joel Lewis. When Lewis died, Whistler did not offer his job to the very competent, but irascible, Boyden. Instead, he asked his former assistant on the coastal railroad to join him in the new city on the Merrimack. James Francis said that he arrived in Lowell on November 22, "taking the place of Joel Lewis as surveyor."[33] Among other things, he was soon responsible for the hydraulic engineering formerly done by Lewis and Boyden. One of his first notebooks, from 1836, shows a young man already studying descriptive geometry and mastering the art of drafting.

Francis's aptitude for technical illustration was exceptional. The Locks & Canals machine shop made quick use of his talents as a draftsman. He apparently helped to disassemble a British locomotive built by Stephenson. Ironically, the Middlesex Canal had delivered this engine of its own destruction. Francis measured and drew all of its parts for the shop's pattern makers and machinists, who were using it as a model for the locomotives they were building for the new Boston and Lowell Railroad, which opened in 1835. That locomotive project may not have lasted very long, for he was soon working on mill and hydraulic engineering. The Boott Cotton Mills were under construction, as was the Eastern Canal, which supplied them with water carried from the lower level of the Pawtucket Canal. A group of distinguished colleagues said that Francis, then only nineteen, was "entrusted with much of the responsible design and oversight" of the Boott mills project for Locks & Canals.[34]

We don't know what Kirk Boott thought about the young assistant engineer. Although he is mentioned briefly in annual financial accounts from 1835, his identification as "our engineer" does not appear in the Directors' Records until September 12, 1840. By that time, Whistler had left Lowell to concentrate on railroad projects, and Francis, at the age of twenty-two, had become the new chief engineer of Locks & Canals.

The canal system as projected on the 1825 plan by George Baldwin was complete when Francis replaced Whistler in 1837; the Eastern Canal had been the last major component. America's economy was entering a depression, and only a few mills were under construction in Lowell. The young engineer could focus on operating the two-level system and keeping it in repair. Working under the scholarly and supportive Whistler, Francis had significantly improved his knowledge of mathematics and engineering. He may even have used the growing engineering library of the Baldwin family in nearby Woburn. The need for waterpower was still increasing, but the limitations of the system would soon put his professional skills to the test.

Expanding the Waterpower, 1836–1847

The town of Lowell became a city in 1836, the same year that marked the opening of the Eastern Canal and the completion of the first phase of waterpower development in the growing industrial center. Lowell now boasted 17,600 residents, and more than 7,500 of them worked in mills, print works, or shops on the Locks & Canals system. There were hundreds of job openings as the first of the new Boott mills went into production. By the end of the year, eight textile corporations (Boott was the ninth) were running 146,128 spindles and 4,667 looms. In all, twenty-five textile mills, two print works, and the Locks & Canals machine shop were drawing power from the Merrimack River.

Even the extended economic depression that began with the Panic of 1837 did not arrest the trajectory of industrial expansion in Lowell. All of the corporations suffered to some extent, but they rode out the troubled times without significant drops in annual production or long-term layoffs. Yards of cloth produced in Lowell went up every year from 1836 to 1841, fell slightly in 1842, and then recovered. The directors of the Boott Cotton Mills continued their building program and increased capitalization by $200,000 during the Panic. In 1839, another large corporation bought mill-powers on the Eastern Canal and began construction next to the Boott site. Because of its position at the junction of rivers, the Massachusetts Cotton Mills discharged into both the Merrimack and the Concord. A smaller corporation, the Prescott Manufacturing

Company, built mills on adjacent land near the Lower Locks in 1845 but was absorbed by the Massachusetts complex within three years. Millpowers leased by the Prescott finally put the two-level system in rough balance. Large flows of water were no longer wasted from either level except in unusual situations. In theory, if not in practice, the lower-level mills used approximately the same amount of water as the upper-level mills.

Locks & Canals continued to build most of the new factory buildings and worker housing, installed waterwheels and power transmission systems, and supplied production machinery. The machine shop, which had been part of the canal company since the corporate reorganization of 1825, was a valuable asset. By 1835, it could turn out all the machinery for a mill of five thousand spindles in four months and make a profit of between 20 and 25 percent on the contract. Two main branches of manufacturing had developed in the shop by 1835: one devoted to textile machines and the other to steam locomotives. In addition, the shop made equipment for controlling water, including gate-hoisting mechanisms used in power canals and mill raceways.

The machine shop used water-powered cutting tools to shape both wood and metal, but it had no foundry for casting metal parts until 1840. It did have men, like millwright John Dummer, who made wooden patterns for those parts. The cost and inconvenience of ordering castings from Massachusetts towns like North Chelmsford, Easton, and Bridgewater prompted Locks & Canals to build its own foundry in 1840 near the shop buildings. The foundry drew water from the upper level at the Swamp Locks basin and, like the rest of the shop, discharged into the lower level of the Pawtucket Canal. Foundries needed power for driving furnace bellows and for grinding or buffing castings.

By 1840, the machine shop and the ten textile corporations that were to dominate the social, political, and economic life of Lowell for the rest of the century were all on the canal system. Most of these plants continued to expand production and to use as much waterpower as Locks & Canals would provide. Their managers were not eager to share land and water resources with additional corporations, but they saw plenty of room for improvement in the system and the way it delivered water. Their plans for future growth depended on reliable, inexpensive power.

One way to make more waterpower available during the work-day was to increase the storage capacity behind the dam. Raising the effective height of the dam both deepened the millpond and expanded its area. In 1832, Locks & Canals had tried temporary "flashboards" (planks held against metal pins that projected from the crest of the masonry dam). These were removed when workers added two feet of granite headers to the structure in the following year. "Flowage" payments for submerging upstream property indicate that two feet of flashboards were back on the higher dam in 1838. Ice floes or spring floods would knock most of them off each year, but they could be replaced. With flashboards intact, the optimal height of the millpond was thirty-four feet on the Locks & Canals scale.

As a waterpower company that contracted to deliver water at both a specified flow rate and a specified head, or drop, Locks & Canals had to respond to many complaints. If just one company on the upper level took significantly more than its allotted share of the flow, then other mills on that level might not get enough power. Water elevations in canals would begin to fall whenever total demand exceeded supply. When river levels were high, the canal company could raise its head gates and let more water into the system, but this took time and did nothing to stop the abuse by the offending company. During drought conditions, that response was not an option, because the river could not deliver enough water to keep the millpond full all day. Greedy power users could quickly cause a critical situation in the summer or fall. Within a few hours, the entire pond could be drawn down so low that it could not deliver water at the contracted head.

The same upper-level mill that took too much water was, of course, discharging that extra flow into a lower-level canal. There it could raise water levels and cause backwater conditions for other mills on the upper level. Thus, neighboring mills on the upper level could simultaneously find water too low in their headraces and too high in their tailraces. The net loss of head and the resistance of the backwater could make it impossible to run all the production machinery. Since textile production was a continuous, sequential production process, it was no easy matter to shut down part of the machinery.

Offenders on the lower level could also cause water levels to drop if they took too much water. The lower level depended on the discharge from the upper level in this balanced system. It was possible to "waste" additional water into the lower level if the river was high, but this was a costly exercise in terms of both labor and lost revenue. The extra water would not be providing any power on the thirteen-foot drop between levels. The canal company did not like to use thirty feet of head to produce power on just a seventeen-foot fall, but they had to do that if someone was secretly drawing too much water on the lower level.

The lower-level mills, which discharged directly into the Merrimack or Concord rivers, needed extra flow during backwater periods, when river levels were too high. In these conditions, vertical wheels ran with difficulty. Their efficiency dropped, and more cubic feet per second of water were needed to produce the same power. Some mills began installing extra wheels that they only used in backwater. Although there was plenty of water coming down the Merrimack in freshets, every drop used by the mills had to pass through the relatively small Pawtucket Canal. More flow meant higher current in the canal, and that caused trouble.

The height of water drops in a canal when friction with the bottom and sides dissipates part of the energy of the moving liquid. These losses in elevation are called "friction head losses." They grow progressively more serious as current increases. If the containing surfaces of the canal are rough, that adds to the friction. The Lowell canals were built in such a hurry that surface finish was often poor, even jagged where tough ledge had been left in place. They could usually deliver more water than mills had leased (except in dry seasons), but they could not always deliver it at the level that was promised in the same leases. The cross section of the Pawtucket Canal was simply not large enough to handle a substantial increase in flow without excessive current, and that meant giving up several feet of head. One report said that mills sometimes "draw so hard upon the Pawtucket Canal as to cause a great descent in it, whereby the level of water at the mills, is from 2 to 2½ feet below the level of the river at the head of the canal."[1]

Low head was just as detrimental to power production as was low flow. In the formula for power, you multiply flow times head (see sidebar, p. 40). Diminished levels in the canals could force mills to run only part of their machinery. In fact, if the head at a mill dropped more than three feet below the prescribed level, the gates might not function at all on some breast wheels, and production could come to a standstill.

Searches for a radical solution to the friction losses in the Pawtucket Canal began in the late 1830s. Previous efforts to enlarge the cross section, or at least to remove some of the ledges that aggravated friction problems, had accomplished very little. The main problem was that Locks & Canals could not shut the canal down for an extended period of time: the manufacturing corporations would not allow it. Some work was possible at night or on Sundays, when the mills were closed, but extensive modifications to the single feeder could cut off flow to the system and stop mechanized production. Neither the canal company nor its customers wanted to see the Pawtucket Canal drained for months.

At the request of Locks & Canals, consulting engineer James Baldwin produced a conceptual design for another feeder canal in 1839. James was the younger brother of Laommi II and George. Like the latter, he had not gone to college but had learned engineering by individual study and by working on projects with Laommi II. He had then distinguished himself as a professional engineer during the construction of the Boston and Lowell Railroad. The plan for a new feeder was a collaborative effort involving employees, managers, and investors in the canal company. William Worthen, a Harvard graduate then working under the direction of James Francis, the twenty-four-year-old chief engineer, drew the route of the canal on a large sheet. It would run from the Pawtucket Dam down the south edge of the river and then turn inland to connect with the Western Canal. This was a short, efficient path to the Tremont and Suffolk mills, whose discharge fed the Lawrence mills. If recommended, the new canal could even change the direction of flow through the Western Canal, thus providing much needed water at the Swamp Locks basin and taking part of the load off the overtaxed Pawtucket Canal.

Patrick Tracy Jackson, who had become the treasurer and agent of Locks & Canals, compiled detailed cost estimates and played a major role in the planning process. He looked beyond the hydraulic improvements to consider other economic benefits that the canal would bring: "Our greatest advantage will accrue from the increased value of our real estate, & from the profitable employment of our machine shop for some years to come."[2]

One undated plan from this period shows the faint outline of additional power canals and mills that could be supplied with water from the new feeder canal. This two-level arrangement is sketched on undeveloped land at the bend of the river, northwest of the existing Lawrence and Suffolk mills. Jackson said in 1839 that "we shall also be enabled to furnish water for new mills at a place where we have a large lot of vacant land."[3] As it turned out, those mills were never built.

Locks & Canals set up a special committee to investigate "the subject of a new canal" and to evaluate the plans. One of the key objectives was to reduce the demands on the Pawtucket Canal and thus eliminate the problem of excessive friction head losses in that small channel. Another goal was to increase the number of "constant" millpowers that could be leased. By Jackson's estimate, an increase of at least thirty millpowers should result from all the steps he suggested. The committee members agreed that additional power could be gained by building the new canal, stopping leaks in the dam, and doing a better job of ponding water at night; but they wanted more than unsubstantiated estimates of the river's potential before they would support "any considerable expenditure."[4]

In what was probably his first opportunity to carry out a major hydrological investigation at the direction of the Proprietors of Locks and Canals, Francis calculated the lowest flow expected in the "natural run" of the Merrimack River. Nature cooperated by providing a severe drought in the summer of 1840, during which he could conduct his experiments. Using Uriah Boyden's earlier surveys of the exact area of the millpond, Francis treated it as a giant basin in which he could briefly trap all the downstream flow. To find the flow rate, he split the eighteen-mile pond up into manageable sections and measured the exact rise of the pond over the area of each section during intervals of time when the mills were not running. After determining the volume of water gained (rise times area) in multi-hour periods on different days, Francis reported that the lowest measured flow was 1,488.63 cubic feet per second. The committee members were aware that "occasions of extreme scarcity" might result in even lower flow, but they had enough confidence in these experiments to make projections about the power that should be available year round. With this information it would not be necessary to "make quite so liberal a deduction as formerly, for the unknown and the contingent."[5]

Francis had calculated that approximately twenty-seven new millpowers could be supplied if all the mills would stop their wheels at night and let Locks & Canals pond water behind the dam for nine out of twenty-four hours. The committee was more conservative in its predictions of the potential gains if ponding was instituted and the new canal built. Members argued that it would be safer to lease only twenty more millpowers. Although the committee's report recognized that "a much larger additional power" would probably be available eight to ten months of the year, the committee based its arguments on the gain in "constant" powers and treated any seasonal surplus as a "collateral advantage." The canal company had not yet worked out a way to profit from extra water that could not be guaranteed year round. The value of that water would "depend on means of reimbursement."[6]

Jackson had already estimated that the new canal could be built for $250,000. The committee used his figure to argue that "the cost of the new Canal, both principal and interest, will be so nearly reimbursed, if not entirely, by the sale, within a reasonable time, of twenty new millpowers, as to justify the expense of the undertaking, considering the other benefits to be derived from it."[7] They expected that Locks & Canals would profit indirectly from sales of machinery and land and that the value of its entire canal system would go up sharply.

The plan depended on each corporation accepting a fifteen-hour daily limitation on power generation, something that was not in the original leases. As both Jackson and Francis knew, the only way to lease twenty more millpowers was to seal leaks in the dam, shut the head gates into the canals at night, and let the river refill the partially depleted pond before work began again in the morning. Only then would the natural flow that he had measured be sufficient.

In 1841, Jackson reported to the directors of Locks & Canals "that the Merrimack Company, will not make such an agreement as to the use of the Mill Power sold to them, as will enable this Company to hold up water in the night time in such a manner as would justify us in building a New Canal."[8] In the face of this objection by the largest customer on the system, the directors shelved plans for a new canal.

Despite the failure to get a ponding agreement, it still made sense to sell some additional millpowers because companies were already using them without compensating Locks & Canals. The critical question was, How much water did each corporation actually need for its normal operations? Locks & Canals had asked James Baldwin to measure the water consumption by the mills in 1839, but his efforts to do so in their tailraces failed because of the high currents in those narrow channels.

Once Francis had completed his experiments on the summer flow of the river in 1840, the directors of Locks & Canals had a better sense of the potential power available but still no reliable data on consumption. They sought answers by employing a commission of three engineers, including Baldwin, to measure flow going to the mills in 1841 (for techniques, see chap. 4). The Merrimack Manufacturing Company was, as expected, the principal offender. In addition to the ten millpowers it leased, it was taking almost six and one-half powers without paying for them. According to the tests, other corporations were also drawing more water than their leases allowed.

Locks & Canals authorized a new set of leases in 1843. The immediate goals were to make the leases reflect reality and to end the regular consumption of water without payment. Jackson leased additional waterpower to the companies on the system for a negotiated fee of $4 per spindle. The canal company agreed to provide even more water than was already in use (with or without its permission). Its action added a total of $20\frac{5}{30}$ millpowers to the $66\frac{20}{30}$ already leased. An additional lease of $4\frac{18}{30}$ for the new Prescott Manufacturing Company (soon part of the Massachusetts Cotton Mills) brought the total to $91\frac{13}{30}$ in 1845 (see table 5.1).

Supplying waterpower had been the primary function of the canal system since Boott, Jackson, and their associates took over the canal company in 1822. Transportation through the waterways, while still important for Lowell's sawmills and helpful for the textile firms, was no longer a high priority by the late 1830s. Nor were tolls a source of much income. Once the railroad linked Lowell to Boston in 1835, its speed and year-round reliability made steam trains the best way to ship out fabric and bring in raw cotton or wool. Sending a few shipments down the Middlesex Canal was helpful to keep railroad rates reasonable, but the trains were clearly winning the competition for Lowell's trade. The directors of that failing canal began liquidating assets in 1844, and they closed it for good in 1853. Corporations did continue to use Lowell's power canals to carry heavy construction materials such as stone and brick when new mills were in construction.

A special set of locks built in the Western Canal in 1833 was the last one added to the system. Two lock chambers linked the mills of the Lawrence Manufacturing Company (under construction on the lower level) to the upper-level canals and, through them, to the river and the Middlesex Canal. Every mill complex was then reachable by boat. In that year, Locks & Canals took in $10,870 in tolls, a figure that partly reflected the demands of mill-building projects. In 1834, the total was $9,158.96, but the following year saw the opening of the railroad, which cut toll receipts almost in half. In 1838, the annual return from tolls was only $2,563.79.

Within a few years after a rail spur had reached the Lawrence mill yard, Locks & Canals closed off the locks on the Western Canal. Above the Swamp Locks, however, the Pawtucket Canal still handled substantial traffic. For decades, lumbermen had brought logs down the Merrimack in tightly bound rafts, using locks to get around the worst drops. After 1841, Fiske and Norcross, and other sawmill operators, began spring drives of free logs, sending them over the northern falls in high water and then trapping them behind booms in the Lowell millpond. There drivers sorted them by company trademark, formed them into rafts, and floated them into the Pawtucket Canal. The charter of the canal company required it to operate locks for rafts twenty-five feet wide.

Repairing locks was expensive, and costs went up with width. The locks built in the 1820s had decayed badly by 1839, when major repairs commenced at the Guard Locks. As they rebuilt the twenty-five-foot chamber at the Guard Locks, which allowed rafts to reach sawmills on the upper Pawtucket Canal, the company's directors were simultaneously seeking permission from the legislature to change the locks that served the lower level. Narrower chambers would be easier to maintain, take less time to fill, and waste a smaller amount of water, which could be used for power generation. The legislators, who must have recognized that few rafts were continuing downstream to the sea, bowed to the wishes of the influential men who controlled Locks & Canals and authorized a width of only twelve feet. Francis then modified the stone walls at the gates and used sheathed wooden framing to narrow the chambers.

Francis had to deal with transportation over the canals as well as through them. In effect, the canal system divided the city into islands, creating physical barriers that could only be crossed at certain points. These points were where bridges spanned the waterways. The canals were corporate property, designed primarily for power and, to a lesser extent, for the movement of materials. Locks & Canals had to build and maintain the bridges that enabled pre-existing public roads to pass over its canals. The corporations also needed some pedestrian bridges so that workers could travel between the mills and their housing without undue delay. The cost-conscious canal company created no more bridges than necessary and may have skimped on railings across or near those structures.

A dramatic newspaper account of a "Melancholy Accident and death by drowning" in the spring of 1840 suggested that railings were inadequate or missing at some bridges and other places where people passed close to the canals. On a Saturday evening, Miss Lucretia Ricker was walking from her boarding house "on the Boott Corporation to visit her friends on the Merrimack." The Merrimack Canal stood between the properties of the two corporations, and "in attempting to cross the canal she missed the pedestrian bridge and fell headlong into the water." Although people heard her screams, "the rapid current of water carried her over the wasteway into the Merrimack River before any assistance could be rendered her."[9]

The newspaper reported that this was the second person who had "drowned within a few months by accidentally walking into the canals of this city." There were "several places . . . where those who are but little acquainted with their localities would be very likely to meet with such an accident. It is manifestly the duty of *somebody* to erect suitable railings at these exposed places so that such fatal accidents may be avoided." The editors appended a pointed note saying that Locks & Canals owned the "canal into which Miss. R. fell."[10]

Maintaining railings was just one of the diverse responsibilities for a corporation that seemed to have too many things on its plate. The directors of Locks & Canals began to consider limiting the company's focus by the early 1840s. Their machine shop had been the best source of income since it began in Lowell. In 1838, the shop accounted for 81 percent of the total operating profits of the corporation, but the market for textile machinery was becoming more competitive, and continuing success as a builder of locomotives was not assured. A single share of stock in Locks & Canals that sold for $500 in 1825 was worth $1,820 in 1836. The depression of 1837, however, eroded confidence. By 1843 a share was down to only $700. This decline in value was not reflective of the company's record for substantial dividends: An investor who bought a share in 1825 and held it for nineteen years would have received $1,422.50 in dividends, which was equivalent to a return of almost 15 percent per year.[11]

There was increasing concern over future prospects for the shop's products. How much larger could the booming textile industry grow? And even if limits were not yet in sight, would the shop in Lowell get enough of the new business? Locks & Canals now had serious competition from other producers of textile machinery. As early as 1841, the directors of the canal company were considering the possible sale of the machine shop. Nothing came of this until 1845, when a group of Boston Associates, including Abbott Lawrence and William Appleton, agreed to pay Locks & Canals $175,000 for the plant. The Lowell Machine Shop became a separate corporation and had to rent its waterpower like any other manufacturer on the system.

Giving up the shop was a key step in the reorganization of Locks & Canals. Everything changed between 1845 and 1846. The canal company sold not only its manufacturing plant but also most of its real estate holdings in Lowell. The textile corporations on the system bought $600,000 worth of property, and the city acquired large parcels for two public parks (North Commons and South Commons). Many other buildings and vacant lots went to private individuals or businesses by auction. Locks & Canals still held a great deal of land, but most of it was directly related to the operation or maintenance of the canal system.

The directors and stockholders agreed to end public ownership of the corporation. The manufacturers on the system, in effect, bought the company on which they depended for waterpower. Accountants calculated the total assets of Locks & Canals, including proceeds from recent sales and the value of all remaining property. Stockholders got $1,582 per share as a "liquidation dividend" for their holdings in the canal company. This sum, plus the dividends paid since 1825, added up to an average annual return of 24 percent for the original investors. The venture had been incredibly successful.[12]

Ten textile corporations and the newly incorporated Lowell Machine Shop were now the owners of the reorganized Proprietors of the Locks and Canals on Merrimack River. The subscription of stock by each corporation was proportional to its existing share of the total allotted millpowers on the system. Thus the Merrimack Manufacturing Company, which leased the most waterpower, became the largest holder of shares in the new Locks & Canals. This change in ownership was a way to protect the interests of the waterpower users. Those who had the greatest need for power now had the most influence over the system that delivered it.

At one time the issue of who managed the waterpower had not seemed so important. The stock in Locks & Canals had been controlled by the same men who set up the first textile corporations in Lowell. That connection had grown more tenuous over time, as more people acquired shares. Corporate leaders apparently feared that an increasingly independent canal company could no longer be trusted to protect their interests or make the best use of the water resources that had brought them to Lowell. In Francis's words, they were afraid that "the use of the water power might become too much extended."[13] By taking over Locks & Canals in 1845, the manufacturing corporations could decide how that canal company operated and what improvements it would make to gain additional power. They could also continue to keep small firms from gaining access to any of the water.

Since 1822, the agents who supervised the daily operations of Locks & Canals had come from the elite ranks of the major investing families. They were men with a financial stake in Lowell's textile industry. Most of them had been closely connected with at least one of the corporations on the system. This had sometimes created real or perceived conflicts of interest, which may have been a factor in spurring the change of ownership in 1845. Professional engineers like Whistler and Francis worked for the agent of Locks & Canals and had limited opportunities to make their own arguments before the directors. That situation changed in 1845 as the waterpower users took direct control of the canal company and reconstituted its management. The treasurers of the stockholding corporations became the directors, and they appointed a new agent with no ties to any of their businesses.

The treasurers' choice was Francis, the engineer whose objectivity, integrity, and growing technical competence had been on prominent display in Lowell since his arrival in 1834. This was a man who would respect the rights and enforce the obligations of every corporation. He could be trusted to present accurate information and offer honest opinions. The directors knew that managing the canal system was a very difficult undertaking. Francis became both special agent and chief engineer, with a salary of $2,000 per year, on September 27, 1845.

Almost immediately, Locks & Canals came very close to losing the services of its newly promoted agent. In December 1845, Francis received a tempting "proposition" from Charles Storrow, agent and treasurer of the Essex Company, which was developing a new industrial city on the Merrimack. Francis wrote to his employers, saying "I am not at all anxious to leave the service of the Locks & Canals provided my situation be made as advantageous as that offered me by the Essex Company."[14] The directors acted quickly to

James B. Francis in middle age. He was only twenty-two years old when he became chief engineer of Locks & Canals. Courtesy of the Center for Lowell History, University of Massachusetts Lowell.

keep Francis in Lowell. He was about to embark on ambitious projects that would make the city famous for achievements in civil and hydraulic engineering.

Even as agent, Francis never had the social status and wealth of the Boston Associates, but he shared one of the characteristic attributes of their group: a love of gardens and urban green space. He was as fond of growing plants and trees as he was of designing great works of engineering. Francis was born in an English cottage surrounded by a garden. After his marriage in 1837, all of his homes had plots for cultivation. Worthen said that Francis gave his first Lowell garden "personal attention with his usual persistence" and that it was "noted for its vegetables, fruits, and flowers."[15]

Francis was a member of both the Boston Society of Civil Engineers and the Boston Society of Natural History. Although membership in the more exclusive Massachusetts Horticultural Society, which included many prominent Bostonians with holdings in Lowell corporations, may have been beyond his social position, he rubbed elbows with textile magnates whose personal gardens were wonders to behold. As a planner and designer, he personified the "green" engineer, who sought a rational synthesis of the natural world and the built environment. He worked tirelessly to beautify land owned by Locks & Canals and to make it accessible to the public. In these efforts, he was following a tradition of corporate landscaping already well established by the founders of Lowell.

A contributor to the *Lowell Offering* in 1841 was impressed by the effects of the corporate planting programs: "Within a short time, shade trees have been placed around the different corporations; and along some of the canals, even double rows of trees have been placed, forming a cool and delightful retreat." She fondly recalled the "many pleasant hours" she had spent "beneath their shade."[16]

Locks & Canals had put parallel lines of trees between Dutton Street and the canal, thus forming one of those "double rows." People had used the strip of land next to the water as one of the community's unofficial promenades since the late 1820s. An 1848 view shows several well-dressed men and women enjoying that allée formed by a clearly defined double row of stately trees. The slow-growing elms were then mature enough to provide softening elements of natural beauty and grace in a landscape devoted to industrial production. Under Francis's direction, the canal company would create more of these tree-shaded walks, or "malls."

Dutton Street, with its mall and rail spur, in 1848. Boardinghouses are at the left, and the Merrimack Canal is to the right, beyond the Dutton Mall. People are promenading in the shade of the trees. O. Pelton, delineator. Courtesy of the Lowell Historical Society.

Employees of Locks & Canals applied similar landscaping treatments to the street and canal corridor south of Market Street. New trees were just taking root when the Boston and Lowell Railroad built the first of what would be a double line of tracks along the canal in 1835. The main line ended at a station on Merrimack Street, but a spur (with horse-drawn cars) soon extended down the Dutton mall to the Merrimack mills.

In spite of the rail intrusion, a lovely mall survived north of Merrimack Street. In 1845, mill operative Josephine Baker described the younger of the two lines of elms, the row along the canal: "To the right was the canal, neatly walled-up and enclosed with a white fence near which a row of trees had been set, that flourished and gave promise in time to shade the railway that ran beneath to the factory yard"[17]

Walkers venturing beyond the developed malls of the Merrimack Canal could follow it upstream to the Pawtucket Canal and sites outside the urban core, such as the falls on the Merrimack. Lucy Larcom, who became a famous writer, recalled strolls in the 1830s along the "old canal path, to explore the mysteries of the Guard-Locks." She said that "there were miles of winding canal, that on holidays tempted young feet into long, sunny wanderings through green pastures." Worthen commented that walks to the Guard Locks "in pleasant company" were "the thing" in his day.[18]

In 1840, the only block along the Merrimack Canal that lacked formal landscaping was immediately south of the train station. That situation changed quickly when Kirk Boott's younger brother, William, became agent of Locks & Canals. He designed the beautiful "Shattuck Mall" for the site, giving it the richest treatment of all the company's planted areas. In William's planting list for the three-quarter-acre mall, historical architect Charles Parrott has identified 115 trees of 43 species and 88 shrubs of 38 species. For a few years, Locks & Canals took care to protect this landscape from encroachment by new technology. The company treasurer told Mr. George in 1845 that he could not "have the right to erect his telegraph in Shattuck Mall."[19]

North of Merrimack Street, the rapid pace of urban development stimulated popular support for the preservation of green space. In 1844, several residents of Anne Street took steps to protect their landscaped surroundings on the east bank of the Merrimack Canal. They persuaded Locks & Canals to grant them trusteeship over the strip of land at the canal's edge so that they could keep it "as ornamental ground forever." The land so designated was "dedicated and set apart by the grantors for the purpose of beautifying and ventilating the City."[20] The conveyance stated that the land was to be devoted to trees, shrubs, grasses, and other plantings. This was to assure that Anne Street residents would continue to receive "the advantages of light, air and prospect that they now enjoy."[21] Locks & Canals, however, carefully retained its right to widen the canal. Hydraulic efficiency and the power demands of the mills came first.

Although the Anne Street park remained a restricted enclave, fenced off at the street line (probably to retain the unfenced waterside necessary for breaking up ice), it did complete the landscaped enclosure of the canal north of Merrimack Street. This tree-lined water allée had an equally graceful architectural perimeter. A harmonious group of residential, institutional, commercial, and industrial buildings enclosed the greenway with a unified and balanced edge. This handsome corridor figures among the significant explorations of urban design in the nineteenth century. Readily accessible and open to multiple levels of society, it tapped social democratic impulses that were only beginning to be explored by America's nascent park movement. Here the boardinghouses of mill operatives faced the homes of the gentry across a shaded canal in the center of a busy city.

The parklike surroundings of the Merrimack Canal and its flanking streets provided a safe and attractive place for people to walk in the company of their fellow citizens or to pause for periods of personal contemplation. Although canals may have been relegated to back-alley status in some places, Lowell celebrated its raison d'être with a delightful greenway that became the most popular gathering place in the city.

Traditionally, promenading has meant walking for pleasure in public. This venerable European practice came early to America but gained increasing social importance here in the nineteenth century. The landscaping along the Merrimack Canal attracted promenaders, particularly on Sundays and holidays, when the mills ceased work. Many of the so-called Lowell mill girls took part in this outdoor activity. Nineteenth-century social reformers, including numerous advocates of urban parks, argued that walking about in the fresh air was a restorative exercise that helped keep workers healthy. They also stressed the importance of bringing people from different classes together and inspiring urbane, cultivated behavior through appropriate example. Industrialists who controlled large manufacturing firms at Lowell saw clear benefits in keeping workers satisfied with their lives, proud of their community, and in good health. These investors also recognized the public relations value of Lowell's image as a model industrial city, where young women could find both productive employment and cultural uplift.

It was common for female workers to save several hours each Sunday, their one day of leisure, for promenading in their best dresses and hats. Many set aside part of their wages earned in the mills for the purchases of stylish clothing. Some family members resented this conspicuous consumption. Fathers, in particular, were often taken aback by the "airs" of their newly independent daughters in Lowell. There was also, as one would expect, a positive response to conspicuous displays of clothing in a city built on textiles and eager to present a cultivated appearance. John Greenleaf Whittier, who worked as an editor in Lowell before gaining fame as a writer, approved of the female operatives dressing up and promenading in the city. He said that following the afternoon service on Sunday, the streets blossomed as if the flowers in a garden "should take it into their heads to promenade for exercise. Thousands swarm forth, who during week days are confined to the mills." To a city with such a large workforce, "the weekly respite from monotonous in-door toil, afforded by the first day of the week was particularly grateful."[22]

Some members of the growing female labor reform movement

had a different response to working women who dressed well in public while accepting speed-ups and low wages in the mills. Promenading with expensive clothing actually became an issue in the factory debates of the 1840s. In 1848, a contributor to a newspaper called *The Protest* said that the "Cotton Lords" used such behavior by mill operatives to prove that existing wages were too high: industrialists could claim that "the girl who works in the mill, *the factory girl,* dresses as well as our daughters."[23]

There was also an ethnic dimension to some of the criticism of operatives who promenaded. Those who walked about on Sunday after church could see Irish immigrants working in the power canals. The only daylight hours when water could be shut off for cleaning or repairing a canal without halting textile production were on Sunday. A satirical booklet called *Easy Catechism for Elastic Consciences,* written for the "Sabbath-Labor-Christians of Lowell" in 1847, noted the unfairness of a system that relegated Irish workers to making "heave offerings" with their shovels from the beds of the canals, giving up religious services as well as "all the pleasures and fashionable amusements" that others enjoyed on the normal day of rest.[24] Whittier agreed with those who found this treatment of the Catholic Irish by Protestant managers hypocritical. He noted with sarcasm: "True, the Sabbath is holy, but the canals must be repaired."[25]

The criticism may have had some effect on the directors of Locks & Canals, who voted in 1849 "that the Agent be directed to take such measures in drawing off the canals as shall prevent as far as practicable the necessity of working on the Sabbath." Less visible night crews took over some of the cleaning tasks and repairs, but work on Sundays did not end. When fifty men were employed "in shoveling & wheeling" sediments from the canal by the Boott Cotton Mills in 1853, Francis noted that "all the men who worked Sunday were paid 50c per day extra—owing to the unusual difficulty of obtaining men."[26]

Promenading along the canals did not eliminate social barriers and discrimination based on class, ethnicity, and occupation, but it did bring people of widely varying backgrounds together in public places. In Lowell, it was a sociable activity that also satisfied common needs for contact with nature. The distinctions between public and private space blurred as Locks & Canals encouraged some recreational use of its property or sold parcels for parks.

The canal company let Tremont and Suffolk Mills buy a parcel of land in 1842 on condition that it be "leveled for a public walk or promenade and planted with trees, with suitable avenues and enclosures . . . for the use of the public, as a mall or promenade."[27] Despite such gestures, Mayor Huntington felt that it was not wise to depend entirely on corporations to provide green space in a city where "we are being hemmed in by walls of brick and mortar." He considered "free, open public grounds" a necessity.[28] Recognizing this demand, Locks & Canals sold the city enough property in 1845 to create two sizeable commons.

Even after the city developed its own parks, landscaping continued to be a high priority for Locks & Canals. The new agent was careful to maintain trees that had already been planted on company property, and he would have his own plans for beautifying future engineering projects. But first Francis had to assure more year-round flow and make a significant addition to the canal system.

Natural flow in the Merrimack River was highly variable over the course of a normal year, and in unusually dry years, it could fall to rates (cubic feet per second) that made it difficult for Locks & Canals to supply the millpowers that each corporation had leased. In 1845, the developers of a new city on the lower Merrimack were growing concerned that their venture would face the same seasonal power shortages.

Samuel Lawrence, an experienced Lowell industrialist, was one of the principal investors in the Essex Company, which was building a substantial dam and power canal at Bodwell's Falls, ten miles downstream of Lowell. He knew that much of the flow of the Merrimack was fed by lakes in New Hampshire. While planning the city of Lawrence (named for Samuel's brother and fellow investor Abbott), he "became alarmed lest the control of those grand reservoirs should be in the hands of parties not in harmony with the mill-owners on the main stream."[29] If the right people had the major lakes, then they could store water in the wet winter and spring and discharge it in the dry summer and fall.

The hydrologic cycle was the basis for waterpower in the Merrimack River valley. This cycle was an engine driven by a combination of gravity and solar energy. Water fell under the pull of gravity as precipitation onto the river's drainage basin. Although some returned directly to the clouds through evaporation or transpiration (from plants), most of it proceeded downhill, either on the surface of the basin or through permeable soils and rock formations. Here again, the force of gravity played a critical role. Eventually most of the subterranean groundflow fed springs that added to the surface flow in streams or discharged directly into the sea. Elevation gave water potential energy as it moved down from the New Hampshire uplands to the tidal level of the lower Merrimack River. Water on the surface gave up that energy of position most dramatically at natural waterfalls and steep rapids or at the dams that men placed in its path. When industrialists in Lowell used a dam and canal system to divert water through waterwheels or turbines, they were capturing part of the river's energy for production. The sun could always be counted on to replace the energy lost, raising water vapor from lakes, streams, wetlands, and oceans and driving the cloud formations that fed the basin. The problem for manufacturers who wanted to operate mills all year was that the engine of the hydrologic cycle did not always run at the same speed.

The idea of using controlled releases from specially created reservoirs or existing lakes to improve flow patterns in dry months was not a new one. In Britain and in continental Europe, artificial storage was sometimes used to augment or replace natural flows for transportation or industrial purposes. Francis at Locks & Canals and Storrow at the Essex Company would also have known about Zachariah Allen's successful application of reservoir storage in nearby Rhode Island, beginning in 1822. As a young entrepreneur building a new mill and dam on the Woonasquatucket River, Allen was distressed to experience an extended drought and low water conditions much worse than he had anticipated. After convincing other mill owners on the same river that preventing future water shortages would require group action and a systemic solution, he led them to incorporate the Woonasquatucket River Company in 1823. The multiple reservoirs they created over the next fifteen years made their river system a much more reliable source of water-power and allowed significant growth of industrial capacity in the drainage basin.

In the fall of 1845, Abbott Lawrence bought the Lake Company, which had important property and water rights at Lake Winnipesaukee in New Hampshire. Key figures from Lowell and Boston backed this move, which led to an aggressive program of acquisition and dam construction lasting through the 1850s. Two corporations, Locks & Canals in Lowell and the Essex Company in Lawrence, became the official owners of the Lake Company. By 1859, they controlled more than one hundred square miles of water surface in three lake and stream systems that fed the Merrimack. Lakes Winnipesaukee, Squam, and Newfound had become giant storage reservoirs serving the needs of the mills in Lowell and Lawrence. Hundreds of smaller lakes and millponds linked to the Merrimack or its tributaries were contributing elements of the system.

Locks & Canals and the Essex Company shared the heavy costs for land, water privileges, channel modifications, and control structures. They also paid for litigation, repairs, maintenance, and the operation of gates and spillways. The Amoskeag Company in Manchester, New Hampshire, turned down an offer to join the Massachusetts companies as an equal partner with a one-third share. That was more than Amoskeag's directors were willing to pay for water that had to pass through Manchester anyway. Prominent Boston Associates were major stockholders in all three of the companies. This dispute demonstrated, once again, that corporations that shared some of the same directors did not always follow the same path. Connections (social, familial, or financial) were not proof against business conflicts. In New England, manufacturing centers like Manchester, that drew heavily on Boston for their capital and on Lowell for their urban model, local pride and traditions of Yankee independence were powerful forces. Not until 1878 did Amoskeag finally become an equal partner in the Lake Company.

Control of the lakes did not come without controversy. Much of the resistance to water management by the Lake Company was over "flowage." The company either had to make compensatory payments for or to buy private land it flooded at certain times in a year. Legal battles and even cases of sabotage or mob action marked the history of the New Hampshire reservoirs after 1845. A satirical account in the journals of Ralph Waldo Emerson reveals that Concord philosopher's dismay over the arrogance of a prototypical Boston-based industrialist: "An American in this ardent climate gets up early some morning & buys a river; & advertises for 12 or 1500 Irishmen; digs a new channel for it, brings it to his mills . . . sends up an engineer into New Hampshire, to see where his water comes from &, after advising with him sends a trusty man of business to buy of all the farmers such mill privileges as will serve him among their waste hill & pasture lots and comes home with great glee announcing that he is now owner of the great Lake Winnipiseosce, as reservoir for his Lowell mills at midsummer."[30]

Another intellectual who lived in Concord was as skeptical as Emerson when he considered changes in the Merrimack watershed made to accommodate industry. Henry David Thoreau remarked that the river "was devoted from the first to the service of manufactures." The Merrimack, "with Squam, and Winnipiseogee, and Newfound, and Massabesic Lakes for its mill-ponds, . . . falls over a succession of natural dams, where it has been offering its *privileges* in vain for ages, until at last the Yankee race came to *improve* them."[31]

In Lowell, as one might expect, there was widespread support for the river system "improvements" undertaken by Locks & Canals and the Essex Company. The confident attitude of local citizens is on display in an 1846 letter from D. Spencer Gilman concerning "the large purchases made in New Hampshire for the purpose of turning water into the Merrimack River." If those things were true, he wondered, "who can set bounds to the future growth of Lowell."[32]

New Hampshire lakes were, indeed, the supplemental water resources that Lowell needed for expanded industrial growth, but the directors of Locks & Canals had to take two more steps before they could lease additional millpowers to their corporate customers. First, they needed agreement by all the corporations to limit their hours of work and allow nightly ponding behind the dam. Second, they needed a more efficient canal to bring water into the system. Lowell still suffered from the physical limitations of the upper Pawtucket Canal, the sole feeder for all the mills that used the Merrimack for power. It was time to build the new canal that they had been discussing since 1839.

Objections by the Merrimack Manufacturing Company to restrictions on its hours of work had been the deal breaker in 1841, when Jackson had made nightly ponding a prerequisite for construction of a second feeder canal. The Merrimack was not the only company concerned: Samuel Lawrence, of the Middlesex Company, claimed that his mills had been running at night with at least two millpowers since the early 1830s. He was still asking to use water for 144 hours per week in 1845.

Francis did not want any water "to run to waste during the night in dry seasons." One of the worst droughts occurred in August of 1845. Gilman observed that "the weather continues dry, no rain of consequence having fallen since my remembrance. I am informed that some of our mills are now running at reduced speed for want of water."[33] This was the kind of weather that made corporate treasurers reconsider their stand on nightly ponding.

The corporations were finally ready to do whatever was necessary to insure more power year round. Opportunity for profit from textile production had never looked better than it did in the mid 1840s. These were boom times in the mills. Cooperation in purchasing land and water rights in New Hampshire showed that manufacturers in both Lowell and Lawrence were committed to long-term improvement of the drainage basin and modification of its natural flow patterns. More water was now certain to be available in dry seasons. That could mean additional millpowers and more income for improving Lowell's inadequate canal system.

By 1846, all the lessees had agreed to accept changes that gave Locks & Canals the right to pond water at least nine hours at night. The directors asked their chief engineer to produce plans and estimates for a canal following the general route that James Baldwin had proposed in 1839. Francis prepared two different plans with characteristic thoroughness. He usually presented at least two alternative designs, with realistic cost estimates, for any substantial undertaking. Sometimes he would also give his opinion, which carried considerable weight with special committees or the entire board. In this case, both proposals had more capacity than Baldwin's original design. Francis estimated that the larger design, one hundred feet wide and fifteen feet deep, would cost $311, 450. The smaller one, with the same length and depth but only eighty-five feet wide, could be constructed for $248, 276. Included in his figures were costs for excavation, dam alterations, masonry walls, a lock, guard gates, and four bridges. He presented his two plans to a special committee made up of Baldwin and Storrow as well as former agent Jackson.

In its report for Locks & Canals, the committee reiterated the problems in the system and explained what a new canal could accomplish. After considering the potential gains from reservoir storage in the New Hampshire lakes and from nightly ponding at Lowell, they recommended building the larger of Francis's two plans. The three experts were choosing a uniform cross section of fifteen hundred square feet, to be cut through difficult terrain. If the capacity of this Northern Canal was added to that of the Pawtucket Canal, they believed that Locks & Canals could bring more than four thousand cubic feet per second (cfs) into the system with much less head loss than it presently experienced. The committee certified the accuracy of Francis's cost estimates but added 20 percent for contingencies. It also pointed out that additional projects would be required for effective distribution of water from the new canal to other parts of the system where it was badly needed. That would mean enlarging part of the Suffolk, or Western, Canal and running a "flume or water course from the New Canal to the Merrimack Canal."[34]

The directors approved Francis's larger and more expensive plan for what became known as the Northern Canal. Their chief engineer split the 4,400-foot canal into three sections to simplify final design, construction management, and contracting. The first section began at Pawtucket Bridge, the southerly span of which was to be replaced by a multiple masonry arch structure over the new canal. The path of the canal required excavation across elevated land, underlain with rock. This cut would create an island between the artificial waterway and the river. The second section was the most difficult of all, because the outer edge of the canal would extend into the natural bed of the river. Here, Francis envisioned the artificial channel's most dramatic feature, a "Great River Wall" as much as thirty-six feet in height and one thousand feet long. It would hold water above the rapids, retaining the head from the millpond as the river continued to drop. The third was an inland section that included a bend, a long straight reach, and a perpendicular junction with the Western Canal. The chief engineer produced general specifications for each section, but his key employees directed construction of only the first two. Boody, Ross, and Company, from Springfield, took on the least difficult third section under a contract with Locks & Canals.

Managing such an enormous project was far from being a one-man show. Francis knew how to delegate authority. The man who was usually on the scene when construction was under way was the company's "superintendent," who took charge of all the outdoor work of the canal company. This was always an individual carefully chosen for his work ethic, technical skill, and endurance. Moses Shattuck was superintendent when Francis came to Lowell. When Shattuck died in 1842, Francis replaced him with Paul Hill, who was his own age and had ample experience with Locks & Canals

The Lowell Canal System in 1848. The Northern Canal (with its Great River Wall) and the underground Moody Street Feeder had just been added to the original system. Mark Howland, delineator. Courtesy of the Historic American Engineering Record.

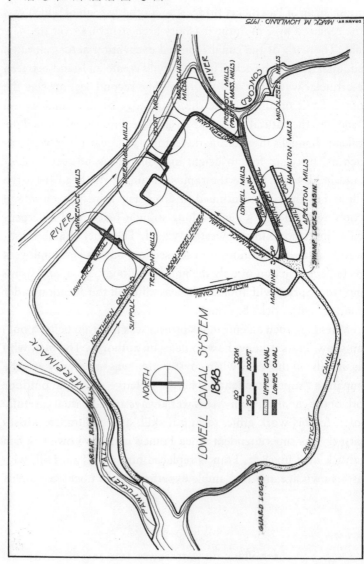

work crews. Hill earned the chief engineer's lasting respect for his outstanding construction management during the Northern Canal project.

Francis left much of the final design, ordering of materials, and scheduling of tasks to his assistant engineers and lower-level supervisors, including Samuel K. Hutchinson, who kept his own notebook and apparently dabbled in an early form of "scientific management." Hutchinson was dealing with ledge removal when he observed that the "daily work of a laboring man is raising ten lb. to the height of 10 ft. in a second & continue this for 10 hours."[35] This gives some sense of the exhausting effort that the canal company expected of its manual laborers.

Payroll records show Locks & Canals expanding its winter workforce of "mechanics and laborers" from a small group of 17 in February 1846, to 128 in March, and to 266 by the end of May,[36] when construction of the Northern Canal began in earnest. Before the summer of 1846 was over, more than 800 men from Locks & Canals, and several hundred more from Boody, Ross, and Company, were hard at work. The canal company kept only 400 to 500 on its payroll during the following winter, but it had close to 1,000 employees in the summer and fall of 1847.

Almost everyone in or near Lowell benefited in some way from the building of the Northern Canal. New jobs were created either directly or indirectly. Orders for materials like stone and gunpowder often went to nearby suppliers. Granite from Dracut or Chelmsford was well known for its high quality and had only to be moved a short distance. Farmers could sell or hire out teams that were not needed every day in the fields but were in high demand for construction: seventy horses and seventy-two oxen reportedly took part. Lowell's shopkeepers and artisans provided a bewildering array of services to the canal company and its employees. Local families recognized the connection between progress on the new canal and the heightened level of economic activity in the city. It was no secret that plans for more textile production were contingent on expansion of the waterpower.

A substantial proportion of the men toiling on the Northern Canal and the Pawtucket Dam in 1846 and 1847 were neither from Lowell nor from surrounding towns. Many were Irish immigrants driven from their homeland by the potato famine. These included transients who moved on after a few months and others who saw long-term opportunities in Lowell. Lasting cultural and economic changes were inevitable as new arrivals decided to stay in the flourishing city, which rose sharply in population, from 20,796 in 1840 to 33,385 in 1850. The Irish community, already well established by 1830, continued to suffer from ethnic and religious discrimination for decades, but jobs were becoming available for Irish women in the textile mills, and construction managers needed hundreds of workers, either native-born or immigrant. While Locks & Canals still favored Yankees for its permanent or more skilled positions, more than half of the temporary labor force was Irish by 1848. For several years, building projects made the canal company the largest employer of male workers in the city.

Most of the work on the new canal required muscle power, with humans, horses, or oxen straining to lift or move heavy loads. Derricks provided mechanical assist for some tasks, such as moving stones into position or lifting earth from excavations. Seven steam engines came into play for pumping water, dredging, and other jobs. A great deal of manual drilling was essential before men could break up hard rock (Dracut Diorite) with wedges or explosives. Although Gilman reported that Locks & Canals had "a little Bullgine up there fixed on wheels, which drills into the rock not lazy,"[37] the typical way of making holes in stone was with a sledge hammer and a hand-held drill rod.

The Northern Canal, like most construction or quarrying sites of the nineteenth century, was a hazardous place to work. Thomas Winslow died instantly when the boom on a derrick fell without warning. A huge rock slipped from the chain of another derrick, killing Michael Reefe, an Irish immigrant who worked for contractor Boody, Ross, and Company. Although the death toll to build the Northern Canal seems to have been no greater than for the erection and expansion of mills in the city during the same period, injuries to workers may have been more frequent. Only the most serious cases received notice in the newspapers or company records. When a temporary building over a steam engine was "blown down," flying timbers struck Thomas Ducy in the head. This unfortunate Irish employee of Locks & Canals was "very badly injured."[38]

Not all fatalities on the Northern Canal project were the result of accidents. The *Lowell Courier* blamed two fatalities on July 8, 1847, on "drinking an excessive quantity of cold water." Once again, the victims were Irishmen, at least one of whom was "newly arrived." Nothing was said in that newspaper about the hot working conditions that were likely the root cause of these unfortunate, and probably unnecessary, deaths.[39]

Blasting to remove rock in the path of the new canal created obvious dangers. Explosives sometimes produced more damage than was intended. They could also detonate prematurely or later than expected. Locks & Canals relied primarily on black powder, a sensitive explosive prone to accidental ignition. One of the two black powder mills in Lowell blew up during the Northern project. A local doctor was experimenting with nitrocellulose powder (appropriately made from cotton) by 1846, but this relatively stable explosive (or propellant) had not yet proven to be cost effective. The invention of dynamite, much safer and more powerful than black powder, was still in the future. Francis did order many thousands of feet of "safety fuse," an improvement over earlier types. Some of this fuse was specially designed for underwater work.

Demolition methods made it risky or uncomfortable to live near the excavations. Locks & Canals paid Prudence Ford "for damages and expenses caused by blasting near her house and injury done to the house." T. L. Lawson wanted "a greater degree of care in blasting." He and his neighbors had "been very much annoyed by the flying of loose stones," and he feared for the safety of his wife and children. He went on to complain that "a stone weighing more than 20 pounds was thrown into my yard over the end of the house, from a vast height—and only a kind providence averted a most serious accident."[40] Another writer in Lowell complained about the noise of explosives, used even on Sundays by the canal company.

What was disruptive for some people was fascinating for others. "Sidewalk superintendents" are not just a modern phenomenon. Hill, the official superintendent, said that "the building of this canal was looked upon as one of the greatest pieces of engineering ever accomplished, and many visitors came to look at the work." Residents of the city were naturally fascinated by the changes taking place at the falls, but this project also drew people from afar. Hill remembered "one noted man after another" arriving to inspect the work. Once he was called over to the carriage of Samuel Lawrence: "When I reached them I found, to my surprise and delight, that Mr. Lawrence wished me to meet Mr. [Daniel] Webster, and to take the party to the bottom of the canal."[41]

After riding down the access ramp, Senator Webster, who was one of the first stockholders in the Merrimack Manufacturing Company, "jumped out of the carriage, and stood gazing long and silently at the great ledge that was being blasted out and at the rough bottom of the falls." Hill then got to see the noted orator in action, as, "he exclaimed, with majestic emphasis, 'The stupendous works of Almighty God are so well adapted to the wants of men!' "[42] Webster's observation fit the prevailing view that such a radical alteration of natural terrain for waterpower was not only necessary but admirable.

If few seemed to object to the new canal on environmental grounds, some tradesmen in the employ of Locks & Canals apparently objected to their wages or working conditions. On July 7, 1847, a supervisor reported that "about 20 masons left the work this morning." This must have been a contentious dispute, for when the masons returned later in the day, "the foreman refused employment" to six of them.[43] The canal company would not tolerate labor protest.

Francis could be rigid and demanding, but he was usually genial in personal interactions and seems to have gotten along very well with his assistants. He demonstrated both compassion for and personal loyalty to superintendent Hill. Since the chief engineer considered him "in many respects the most valuable man in my employment," some self interest may also have been involved. When Hill experienced repeated attacks of "bilious colic" and got no relief from local doctors, Francis sought "the best advice the country affords." He took Hill to Boston to see an outstanding physician.[44]

While the excavations and the building of canal walls were progressing, Francis came out to the falls frequently. Critical parts of his overall project were planned for the entrance to the Northern Canal. There he needed a series of interconnected structures and machinery to regulate the admission of water, to open the artificial channel for occasional boat traffic, and to protect the canal system from floods. The key feature was a gatehouse that would shelter the hoisting machinery for heavy sluice gates. A major section of the dam that Locks & Canals had constructed between 1830 and 1833 stood across the direct path to the new canal and would have to be removed. Francis intended to rebuild that part along the diagonal line of the 1826 dam, which had followed the earlier path of a wing dam for the Bowers sawmill. It was a logical design, which took advantage of rock formations that could become part of the foundation. The reconfigured dam would funnel water smoothly toward the canal entrance during working hours. With the help of flashboards, it would hold back water at night for most of the year. When the river was at flood stage, rampaging torrents would tear out the disposable boards and spill over the long crest of the dam, sparing Francis's masonry gatehouse from their full force.

Keeping the river away during the construction at the entrance was no easy task. For more than a year, temporary timber cofferdams and steam-operated pumps provided essential protection for the unfinished work. Francis retained the obsolete 1830–33 section of the dam as long as he could, but once his men started to remove that heavy masonry and wood structure, only a single cofferdam stood between them and the elevated millpond. The *Lowell Courier* claimed that the temporary barrier was "a most excellent one—said to be the best in the country."[45] Unfortunately, the Merrimack failed to cooperate with Francis's plans. A freshet assaulted the cofferdam at the falls in June 1847. Despite heroic efforts to save it with loads of straw and sticks, the river forced its way through and did significant damage to work in progress.

Storrow, working downstream on a dam for the Essex Company, suffered similar harm from the same freshet that hit Lowell. His cofferdam was also swept away, but he went on to complete one of the greatest masonry dams of the century. Unlike Lowell, where even a relatively low dam was sufficient to "improve" a natural waterfall and rapids with thirty feet of drop, Lawrence needed a tall dam to amplify the existing drop of five feet at Bodwell's Falls. Francis was part of a commission that set the maximum elevation of the Lawrence Dam at thirty-two feet. He made sure that the dam would raise the river no higher than the foot of Hunt's Falls, just south of the confluence of the Merrimack and Concord. There would be no possibility of backwater in Lowell caused by the pond of the Lawrence Dam. Locks & Canals might have damaged the waterpower in Nashua with its own dam in 1833, but it was not going to let the Essex Company do the same thing. Francis must have been planning the removal or lowering of Hunt's Falls to create a greater net drop in Lowell.

The summer freshet of 1847 strengthened Francis's resolve to protect the entrance to the Northern Canal and may have prompted his research into the history of floods on the Merrimack River (see chap. 5). When completed, the heavy masonry abutments and piers of the new bridge across the canal backed up his main line of defense. Adjacent to the tunnel that sheltered the narrow lock chamber were ten guard gates, each at a sluice formed in the cut granite foundation of the gatehouse. The dam joined the gatehouse at its outer end, and both were solidly anchored to the tip of the "island" dividing the canal from the lower river.

In further deference to the river's fearsome power, Francis put a curved corner on his Pawtucket Gatehouse. If ice floes or flood-borne trees hit the exposed portion of the brick building sitting on its stone foundation above the dam, he wanted them to glance off

Pawtucket Dam and Gatehouse in high-water conditions, ca. 1985. There are ten sluice gates to admit water to the Northern Canal. A lock for boat traffic is to the right.

without damaging the masonry. This example of functional stream-lining is still intact after more than a few of the impacts he predicted.

Locks & Canals was building structures that would last. Joseph Frizell, who joined the company in 1857, said that "massive solidity was the characteristic of all Mr. Francis's designs."[46] The chief engineer took exceptional care with the outer edge of the Northern Canal. This was the barrier between the artificial channel and the river. It also acted as an extended dam, because it had to resist the pressure of the upper-level water being carried toward the mills. Immediately downstream of the Pawtucket Gatehouse, Francis used an impervious concrete "spiling wall," encased within a battered (sloping) rubble stone wall. Then there was another spiling wall, this one of rubble laid in cement. It was almost entirely below ground on the island, protected by puddled clay backfill, earth, and

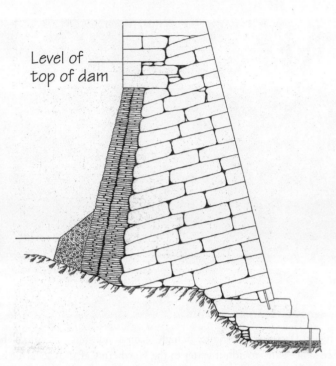

Level of top of dam

The Great River Wall in cross section. The river is to the right, the Northern Canal to the left. Small stones set in cement form a waterproof lining, with a lower band of concrete. Reprinted from *Tenth Census of the United States* (1885), 16: 79.

two stone retaining walls. These mostly hidden spiling walls were intended to minimize leakage, but the most spectacular feature was out in the open for all to see: Francis's Great River Wall, rising at a steep angle from the bed of the Merrimack.

The Great River Wall towered thirty-six feet above the solid bedrock. Granite blocks protected it from damage by the river, while much smaller stones combined with cement made it waterproof on the canal side. The *Lowell Daily Courier* called it "a most splendid piece of workmanship, built with great strength."[47]

At the end of that imposing wall, the Northern Canal turned sharply and headed toward the Suffolk and Tremont mills, situated on either side of the Western Canal. Since 1828, water had flowed from the Swamp Locks basin to those factories. Now the Northern Canal would give them a better supply of water, with several feet of additional head; and it would change the whole function of the Western Canal, reversing its course and sending water flowing down it into the Swamp Locks basin, where it would help the Pawtucket Canal supply all the other mills on the upper level. A final piece of the Northern Canal project was still under construction in 1847, a feature unnecessary for the opening but eagerly anticipated at the Merrimack mills. Francis was building an underground feeder—in effect, a highly efficient shortcut—from the Western Canal to the Merrimack Canal.

The opening of the Northern Canal on Thanksgiving Day in 1847 was a dramatic event that Hill still remembered clearly after almost half a century. He said that "the water was allowed to flow into the canal by blowing up the coffer-dam with two barrels of Whipple's Best Gunpowder."[48] Explosives for this and other dam and canal projects came from a local firm that had been grinding black powder in a Concord River mill before the Boston Associates came to East Chelmsford. In this way, waterpower from the smaller of Lowell's two rivers helped to harness the larger one.

CHAPTER FOUR

Testing the Waters
Scientific Engineering in Lowell

Locks & Canals faced increasingly difficult tasks as the canal system became more complicated and the demands for waterpower increased. The canal company was the first place that manufacturing executives turned to find answers for their engineering problems. James Francis made every effort to keep up with the explosion of technical knowledge in the nineteenth century and in the process became one of America's best-informed engineers. The directors for whom he worked expected him to design structures and machinery with great skill and to apply sophisticated engineering concepts to solve their many problems. His rigorous application of mathematical analysis and precise experimental methods gave engineering in Lowell a scientific tone that was rare in American industrial settings. Yet his publications on hydraulic testing, which brought him international fame, were not highly theoretical. Francis was primarily an engineer, not a scientist; his intention was always to improve engineering practice. He took very seriously his responsibility to give useful advice to the textile executives who leased power and ordered both machinery and factory buildings from Locks & Canals. After the corporate changes of 1845, which gave the textile treasurers direct control of the canal company, he was truly "the Engineer of the Corporations at Lowell."[1]

In any nineteenth-century system of power canals serving multiple enterprises, there were bound to be difficulties distributing the flow of water. A two-level system with a complex layout, like the one in Lowell, was particularly hard to manage. Previous chapters have already discussed the need to keep the system in dynamic equilibrium. Water discharged by mills on the upper level supplied mills on the lower level. Francis had to make sure that all lessees got their fair share of the water. According to Hiram Mills, a graduate of Rensselaer Polytechnic Institute and a former assistant to the chief engineer, "a prominent part of Mr. Francis' duty at Lowell was to distribute water power among the several corporations in accordance with their respective rights; this has called for the execution of many original hydraulic experiments on a large scale."[2] His work in scientific engineering helped him develop reliable procedures for measuring the efficiency of hydraulic prime movers and the amount of water used by the mills.

Information on flow rates was not easy to obtain when Francis became chief engineer of Locks & Canals in 1837. His employers and some of the corporations they served had been struggling with flow measurement for years. Studies of waterwheel performance at a Nashua, New Hampshire, textile mill and at Boston flour mills in the late 1820s relied on a cumbersome method that divided the flow from a waterwheel, during a timed interval, into several equal parts. A special box then captured one of those parts during a timed interval. By measuring the volume of water in the box, a technician could determine the flow rate. Unidentified engineers from Lowell got credit for devising this technique, but it was not suitable for frequent monitoring of flow.

The gauge wheel that Ithamar Beard had used at the Hamilton tests in 1830 was not any more appropriate for regular measurements. Nevertheless, its short-term application provided evidence that mills were drawing more water than they had leased. Consultant James Baldwin was unable to prove the extent of the overdrafts in 1839. He had one more opportunity as part of the commission that studied the canals in 1842. For those tests, he was joined by two other respected engineers (George Whistler and Charles Storrow) and assisted by Francis, who had impressed Patrick Tracy Jackson with the "perfect" method he had used to measure the flow of the entire Merrimack River in 1840.[3]

Despite Jackson's praise for Francis, the directors of Locks & Canals apparently felt that they needed the skills and authority of outside experts to measure the actual water use on the canal system. They were willing to spend generously for a special set of tests using gauge wheels, but they were also thinking about long-term monitoring. If, in the course of their testing, they could find a simple way to measure flow without interfering with the operations of the mills, that would help the canal company keep the system in balance and avoid overdrafts during periods of water shortage.

Commissioners Baldwin and Whistler already had a great deal of experience with the Lowell canal system. Storrow was an excellent choice for the commission because he was perhaps the best-educated American engineer of the period and the author of *A Treatise on Water-Works* (Boston, 1835), the first American book to focus on hydraulics. Storrow had graduated first in his class from Harvard in 1829 and had studied in France for several years at both the Ecole Polytechnique and the Ecole Nationale des Ponts et Chaussées. His Lowell connections were strong: he was the managing engineer for the Boston and Lowell Railroad at the time, and he was related by blood or marriage to some of the prominent Boston families (Appletons, Lowells, Jacksons, and Cabots) who had invested heavily in Lowell's industrial development.

The three commissioners decided not to gauge the water used by each mill building in a complex, but instead to do their measurements "in gross before or after the water enters the establishment."[4] They tried a procedure that had been developed by Pierre DuBuat and further refined by Gaspard de Prony in France. It relied on surface floats and rectangular flumes built in the canals. These wooden structures, which had a consistent cross section, carried all of the water going to a mill complex. Timing the passage of a float over an established distance through the center of a flume gave the velocity of water on the surface. You could then apply the continuity principle suggested by Leonardo and developed as a key formula by Castelli: $Q = VA$, or flow equals velocity times area. It was easy to get the cross-sectional area from the depth of the water and the width of the flume. The problem was that water moved more slowly near the sides and bottom than it did in the middle. The French experts understood this and had found correction coefficients to allow for the difference between surface velocity and mean velocity.

What the commissioners in Lowell needed were specific correction coefficients for surface float measurements in each of their flumes. Under their direction, John Dummer and his workmen built very large gauge wheels across widened, relatively shallow sections of several canals, downstream of the flumes. These wheels, set in close-fitting aprons and turned by the current, in theory could measure all of the water passing through a canal while mills were running. Since the same flow went through both the flume and the gauge wheel, comparison of test results allowed the commissioners to determine the proper correction for a float in the middle of the flume. Locks & Canals then removed the gauge wheel and left the flume in place. After that, a technician had only to calculate the flow in, for instance, the Merrimack Canal flume, using the velocity of a surface float, and then multiply that figure by the correction coefficient for that particular flume (0.814), to get a reduced and more accurate number of cubic feet per second.

The chief engineer of Locks & Canals gained significant stature from his participation in the commission's experiments. Writing much later about Francis's early career, another group of three distinguished professionals disclosed Storrow's role in the discovery of the young man's abilities: "The details of the experiments and measurements of water were entrusted chiefly to Mr. Francis, and so well satisfied was Mr. Storrow with his accuracy, judgment, and skill, that he advised the managers of the Company to rely on him for such work as they might need in the future."[5]

As a result of the measurements in the canals, Locks & Canals negotiated new leases in 1843. This did not, however, solve the problem of monitoring flow. Correction coefficients were not enough to assure consistent accuracy. Surface floats were easy to use, but they did not always move straight down the center of the flumes; and Francis recognized that changes in the current of a canal might alter the relationship between surface velocity and mean velocity.

Francis showed more interest in the water moving through his canals than in the breast wheels that powered all of Lowell's textile mills before 1844. This may have been because those prime movers had been built by Dummer, who was known to deeply resent any interference with his work. Dummer's legendary independence and irritability caused problems for his employers, but his craft skills were so valuable that Locks & Canals could not afford to lose him.

Francis waited until the fall of 1843 to conduct his only serious tests of a breast wheel. As Beard had done in 1830, he used a gauge wheel in the tailrace to measure the water discharged by the prime mover. His subject was a thirteen-foot "Breast Wheel in the Machine Shop of the Proprietors of Locks and Canals."[6] It had an iron shaft and wooden buckets and had apparently been in use since the 1820s. Francis found that its best performance was only slightly better than that of the similar breast wheels tested by Beard in 1830; with no backwater, the machine shop wheel could attain an efficiency (ratio of actual to gross, or potential, horsepower) of about 63 percent.

Testing in backwater, as expected, revealed a serious weakness in the operating characteristics of this breast wheel. Francis included a series of experiments with water at various heights in the wheel pit at the machine shop. When there was just a foot of backwater, the efficiency dropped below 60 percent. By the time backwater had risen to a height of three feet, the wheel was capturing only 37 percent of the potential energy in the water from its headrace. Francis never published the results of this historically significant investigation.

Since backwater was a problem for much, and sometimes most, of the year in Lowell, it may seem surprising that Francis did not do additional investigations of its effects on breast wheels. Perhaps he saw no real solution for the vulnerability of vertical waterwheels to backwater in their wheel pits. Ventilated buckets, essential for running breast wheels in any level of backwater, were already a mature technology. They helped but did not cure the problem. Also helpful in low levels of backwater was simply putting more flow onto the wheels. There were apparently no attempts to resurrect the (probably ineffective) backwater machine of Jacob Perkins, which Francis Cabot Lowell had tried at Waltham. Installation of iron breast wheels with curved buckets at the Prescott Mills in the mid-1840s probably raised operating efficiency a little but would not have produced significant improvement in backwater conditions.

A diary by one of the female operatives at the Middlesex mills in Lowell gives a good sense of how serious a problem backwater could be. Susan Brown wrote on April 15, 1843: "Back water—came out [of the mill] at noon." By the next day the river was in full flood: "Called at Middlesex—back water—did not work. Rainy. Great, long dull day. . . . Went down to see the water."[7]

Breast wheels, which drove all of Lowell's riverside textile mills at that time, could not plough through that heavy backwater. An article titled "The Freshet-Stoppage of the Mills" appeared in the *Lowell Courier* three days later. It said that the Merrimack was almost fifteen feet above its usual height, "and all the mills on the lower level have had to stop work in consequence of back water."

Francis did not have to tolerate the limitations of breast wheels for very long before an alternative technology became available. American artisans who produced small reaction wheels were already demonstrating the potential of turbinelike forms that could run submerged. These horizontal wheels turned at higher speeds than breast or overshot wheels, and some were achieving much better efficiencies than traditional tub wheels. By the 1840s, word of significant French achievements in turbine design was moving across the Atlantic. Ellwood Morris published an article in the *Journal of the Franklin Institute* in 1843 describing his own work with the outward flow turbine of Benoit Fourneyron, who had developed a successful design in France by 1832. Morris also wrote about testing procedures and published a translation of Arthur Morin's important experiments with turbines in 1838.

Uriah Boyden read Morris's articles and began work on his own

Uriah Boyden's outward-flow turbine at Tremont Mills. Water in an iron penstock enters the turbine from above. A flyball governor is to the right. Reprinted from James B. Francis, *Lowell Hydraulic Experiments* (1855), pl. 1.

version of Fourneyron's prime mover. He and Francis recognized the superiority of turbines almost immediately, but it was Boyden who first began making them. In 1844, he designed an outward-flow turbine for the picking house at the Appleton mills in Lowell. Water under pressure entered the center of the wheel from above, through an iron conduit called a penstock, and was directed outward by fixed curved guides. The critical moving part, a runner with curved buckets, or vanes, surrounded the guides. Because the runner reacted to the pressure of the water on its buckets, this class of wheel was called a *reaction turbine*. Having given up most of its energy, the spent water exited the runner at the outer periphery of the wheel with minimal absolute velocity. Francis said that Boyden "introduced several improvements of great value" in this turbine, which produced 75 horsepower and had a tested efficiency of 78 percent, considerably better than the breast wheels in the same complex.[8]

In Lowell, the most important characteristic of reaction turbines may have been their ability to operate underwater. Unlike breast wheels, they were not affected by the drag of backwater in a wheel pit. The only negative effect of backwater on a Boyden-Fourneyron turbine was a reduction of effective head (there was less drop at a dam or falls if a river was in flood). In addition to their relatively high efficiencies and superior performance in backwater, the metal construction of turbines gave them exceptional durability. They were much more compact than breast wheels, required smaller wheel pits, and were less of a hazard. The open wheel pits in Lowell mills (some with linked breast wheels, whose combined bucket length reached eighty feet) were a menace for the unwary.

Francis's self-educated assistant, Joseph Frizell, took a rookie engineer with him to measure a breast wheel that was about to be replaced by a turbine in the winter of 1859. On the way down some ice-covered steps, the new employee slipped and ended up against "one of the great arms of the wheel." Frizell noted that had the wheel "been moving it would have crushed him like an egg, and the water-power of Lawrence would now have a different, though prob-

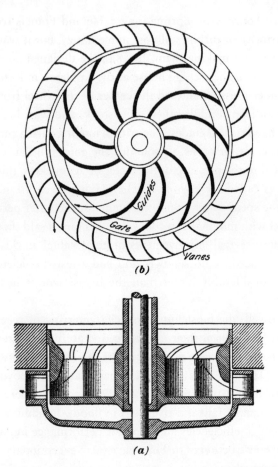

(b)

(a)

Operation of an outward-flow turbine. These section views—*(a)* from
the side and *(b)* from above—show the path of water, first through the
stationary guides and then through the vanes of the moving runner.
Reprinted from *Waterwheels* (Scranton, PA, 1907), no. 40, a correspondence course from the International Textbook Company, fig. 2.

ably not more efficient engineer."[9] The young man who fell was none
other than Hiram Mills, whose later experiments with sand filtra-
tion of polluted water at Lawrence saved thousands of lives.

Once installed, turbines could adapt readily to moderate changes
in head. Their high operating speeds allowed millwrights to cut ex-

penditures on some of the gearing and pulleys formerly used to step up from the five or six revolutions per minute of the breast wheels. This also meant less energy lost to mechanical friction in the transmission of power. Early turbines, usually built on order for a particular customer, were expensive, but astute manufacturers thought they were worth the investment.

With turbines, as with breast wheels, the key consideration in judging performance was their efficiency, a measure of how well they used the potential energy in water. By determining the efficiency of experimental turbine forms, designers could select particular features that gave the best results. Trial and error had long been a part of the millwright tradition, but there was also a theoretical foundation for some of the experimental methods used by Uriah Boyden and Francis. European hydraulicians worked closely with mathematical theories of mechanics and hydraulics and proposed modifications or corrections as experience suggested. Boyden was well aware of European practice as he began to test his own versions of the Fourneyron turbine. Francis sometimes joined him in this investigation. While their relationship was sometimes prickly, it was also remarkably productive. The two men combined the hands-on approach of the American millwright with the quantification and theoretical analysis of the European engineer.

An engineer can determine the efficiency of a turbine if he can accurately measure three things: (1) the head acting on the wheel, (2) the flow rate through it, and (3) the actual power it produces. Multiply (1) times (2) to find the gross power (see sidebar, p. 40). As shown above, the ratio of actual to gross power is the efficiency. Boyden was not only a gifted builder of turbines but also a masterful designer of the measuring devices that were essential to test them.

The head, in this case the vertical distance between water in the headrace and tailrace, was not as easy to measure as one might think. It was not just a simple case of reading height off a graduated scale inserted in the water. Capillary attraction and surface tension draw the water upward at the scale, making it hard to read the level accurately. Boyden designed a special gauge, which used a hook to approach the water surface from below. This tactic took advantages of the same characteristics of water that distorted measurements with an ordinary scale or rule. When the point of the hook reached the surface of the water, a visible bump appeared. The operator then lowered the hook slowly until the bump disappeared, and he read off the correct elevation from a vernier scale linked to the hook. The gauge was accurate to less than one-thousandth of a foot with good lighting and calm water.

Boyden's hook gauge was also a crucial tool when engineers measured flow rate using a weir, a notch or open aperture over which water spills. If you knew the length of a rectangular weir and the head (height of water above its sill), you could apply one of the formulas devised by British or continental scientists to find the flow. Weirs had not been used for flow measurements in Lowell canals because they required a drop in water level, and textile mills could not afford that loss of head.

The weir formula published by Professor John Robison of Scotland had figured prominently in *Tyler v. Wilkinson,* a famous law case of the 1820s that dealt with water rights in Pawtucket, Rhode Island. Boyden knew from his reading of European technical literature that a number of somewhat different weir formulas could be applied to measure flow. He tried several of the available formulas.

Francis also investigated the use of weirs, probably at Boyden's suggestion. Boyden had explained the theoretical foundation of the weir formula, but Francis was uncertain about the exact form that would give the most accurate results. In December 1844, Francis was reading *Experiences Hydraulique* (Paris, 1832), written by Poncelet and Lesbros, and taking notes on the French engineers' weir experiments. By the end of the following year, he had carefully recorded their principal weir formula in his notebook "Experiments and Transactions." At the same time, he mentioned that he was using a slightly different version for some weir situations, based on his own experiments. Boyden was also working with weir measure-

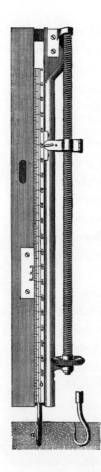

Boyden's hook gauge. This instrument, equipped with a vernier scale for readings to the thousandths of a foot, can measure the head (elevation) of water with impressive accuracy. The hook (shown from two directions at the bottom) approaches the water surface from below. Reprinted from *Hydraulics* (Scranton, PA, 1907), no. 38, a correspondence course from the International Textbook Company, fig. 4.

ment in this period. Modification of the weir formula was under way in Lowell and would continue into the 1850s.

In 1845, Boyden used a weir and a weir formula to measure the discharge or flow leaving the first turbine he built for the Appleton Company. Flow rate and head gave him the gross power. To find the actual, or shaft, power produced by the wheel, he applied a particular type of dynamometer known as a Prony brake, after its French inventor Gaspard de Prony. Once again, his knowledge of engineering developments in Europe gave him an advantage. He may have learned of the brake from one of Ellwood Morris's articles in the *Journal of the Franklin Institute*, but he also read scientific literature in French. Boyden was quick to borrow helpful technology. He could also be harsh in judging the work of others and was adept in finding weaknesses. Most of all, Boyden was a master of incremental improvements.

A year before, when Francis was trying to determine the power produced by the old machine shop wheel, he had needed help to stop sharp jolts and erratic action in his Prony brake. He finally solved the problem by adding a dash-pot, or "hydraulic regulator." Robert Thurston, a professor of mechanical engineering who corresponded with Francis, credited Boyden for the idea and said that it "was probably the first application of this important detail in the use of this brake."[10]

As Boyden worked on turbine design and testing, he was putting together key elements of a hydraulic research laboratory and refining his testing methods. He continued to improve the Boyden-Fourneyron turbine when he made three more wheels for the Appleton Company in 1846. Francis pointed out that Boyden had a financial incentive to make these turbines "as perfect as possible, without much regard to cost." The contract would pay him $1,200 if the turbines achieved an efficiency of 78 percent, but it added a bonus of $400 for every percentage point above that minimum. For these wheels, "the workmanship was of the finest description, and of a delicacy and accuracy altogether unprecedented in constructions of this class."[11]

Each of the new turbines produced approximately 190 horsepower and incorporated numerous modifications by Boyden. He had invented a diffuser to capture kinetic energy still in the water as it left the runner. An ingenious bearing on the vertical shaft suspended the turbine from above. A special flume put the entering water into spiral motion, and curved guides directed it onto the buckets of the runner with minimal shock. Francis helped with these tests and did the calculations necessary to turn observed measurements into the performance figures that determined Boyden's total payment. The data indicated a maximum efficiency of 88 percent. The tests also showed that the diffuser alone accounted for 3 percent of this remarkable improvement in performance. Later improvements in turbine testing and in weir formulas would raise questions about the actual efficiency achieved with these Appleton wheels, but there was no question that Boyden had advanced the art of turbine design.

During the same year, Francis helped Boyden test another turbine at the Merrimack Print Works. The chief engineer carefully recorded the observations in his notebook and sketched parts of Boyden's outward-flow wheel. At one point, Boyden stopped the tests and had the ends of the buckets cut back slightly. After further experiments, he made other alterations and repeated this trial and error process. Calculations of efficiency using data from the careful application of hook gauge, brake, and weir indicated whether Boyden's changes were producing positive results.

Even Boyden's earliest turbines made breast wheels look very bad by comparison, and that must have grated on Dummer, who built the best of those huge prime movers. A biographer said that Dummer "would never look at a turbine wheel; but yet had the curiosity to depute one of his trusty workmen to report to him how it operated." He claimed that the switch from wood to iron wheels ruined Dummer's "vocation" and that, "his usefulness at an end," he retired to a farm.[12] It is a good story, testifying to the painful devaluation of artisanal skills. The reality, however, is more nuanced. Dummer actually made a turbinelike prime mover of his own in 1844, a strange hybrid, made of both wood and iron, that retained some features of a vertical waterwheel. He must have asked Francis to test it, because a drawing and records of experiments with Dummer's wheel have survived. Francis found that it did not perform significantly better than a breast wheel. That disappointment and Boyden's almost simultaneous success may eventually have caused the notoriously temperamental Dummer to retire, but we have to wonder what that master craftsman would have done if his invention had proven to be more efficient than a Boyden turbine.

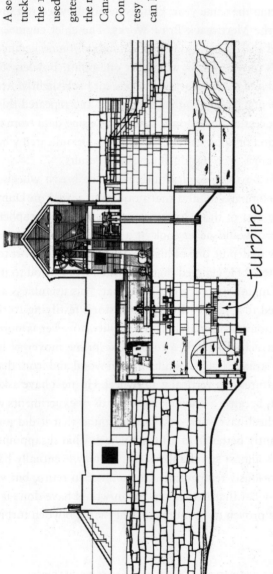

turbine

A section through the Pawtucket Gatehouse, showing
the 1847 Francis turbine
used to open and close
gates. The millpond is to
the right, the Northern
Canal to the left. Michael F.
Conner, delineator. Cour-
tesy of the Historic American Engineering Record.

Francis was also working on a turbine of his own and discussing design concepts with Boyden. Locks & Canals had bought the local rights to use Samuel Howd's 1838 patent on an inward-flow turbine. Francis saw an advantage in the inward-flow, or "centrevent," design, which produced a narrowing of the passages between buckets from entrance to exit. Howd had given up on this approach, but Francis experimented with a simple model that showed promise in 1847. Boyden was present at the tests, in which his hydraulic regulator for the Prony brake and his hook gauge played key roles.

The chief engineer seized an opportunity to build an operational inward-flow turbine later in the same year, when he needed a prime mover for the new Pawtucket Gatehouse at the falls. This was an unusual application in which efficiency was not a major consideration, because the "hoisting wheel" would only be used a few times each day to open or close the heavy gates. An ingenious set of screw drives reduced the force necessary to raise the gates. Francis could afford to take chances with a new form of wheel and could test it to see how well it worked. Unlike the inward-flow model that Francis had just tested, this turbine had guides and floats that were curved. Francis said that his own design was similar to but "essentially different" from Howd's. He admitted that it might be covered by the Howd patent and also mentioned that Poncelet had proposed an inward-flow wheel in 1826.[13]

The handsome brick gatehouse and most of its early equipment is still in place. Francis designed part of the building to function as a research facility, configuring the structure to accommodate special chambers for weir and head measurements and providing versatile spaces for future experiments. It was perhaps the first purpose-built hydraulic laboratory in the United States. Although he never put it to continuous use as a permanent testing facility, Francis returned to the gatehouse time after time to do research. The 1848 tests of his "hoisting wheel" in the building are fully recorded in one of his notebooks but were never published. The experimental turbine achieved efficiencies of more than 77 percent, good results but not as high as the latest Boyden-Fourneyron wheels.

Belt drive from the 1847 Francis turbine in the Pawtucket Gatehouse. The large pulley is on the vertical shaft from the turbine. Gate-hoisting machinery is on the floors above. Jack Boucher, photographer, 1976. Courtesy of the Historic American Engineering Record.

While Francis and Boyden were helping each other with turbine investigations in the late 1840s, Boyden was also seeking patent rights for and profit from his wheel improvements. He was an independent businessman as well as an experimental investigator. Francis said that he "was much surprised" to learn in the spring of 1848 that his colleague had just patented something that the chief engineer considered his own contribution.[14] It is quite possible that Boyden, in this case, took advantage of his working relationship with Francis.

According to a statement by Francis, he had told Boyden about his intension to use a new shape for the buckets of an inward-flow wheel that he was planning for the Boott Mill. The successful designer of outward-flow wheels seemed critical of the idea at the time and gave Francis "not the slightest intimation that he claimed it as his invention." When Francis brought up the subject again during a chance encounter several months later, Boyden "replied that he had an application then pending for patents" for this idea.[15]

Although Boyden had so far produced only outward-flow turbines, he had become as interested as Francis in the inward-flow type. The key improvement that both of them claimed in 1848 had to do with extending the buckets downward as they approached the central axis of the wheel. Because the passages between buckets in an inward-flow wheel narrowed from entrance to exit, water would accelerate in the reduced cross section and leave the runner with too much kinetic energy. Deepening or flaring the buckets alleviated this problem, which Boyden called "choking."[16] Whether this was Boyden's original idea, Francis's original idea, a simultaneous but independent innovation by each man, or a collaborative insight is open to argument. As we have already noted, Francis thought he had conceived the idea on his own, but Boyden made a patent application for a wheel with a similar bucket form. The resulting inward and (slightly) downward movement of water through a runner was an initial step toward true "mixed-flow" and helped lead to the development of smaller, faster, more efficient, and more powerful turbines. It is doubtful if either man understood the significance of this improvement at the time he made it.

When Boyden offered to let Locks & Canals use all of his turbine patents for a price, Francis recognized their value and worked out a mutually acceptable arrangement. The chief engineer and the directors decided not to fight with Boyden over the disputed parts of his patents. In 1849, the directors paid him $16,000 so that they could apply his ideas and sell wheels with his improvements in a defined area around Lowell. They also believed that this payment would encourage him to cooperate in future turbine development by the canal company. Francis was always generous in his published acknowledgments of assistance. Boyden remained a valuable, but often reluctant, advisor. Although he continued to correspond occa-

Francis's inward-flow turbine built for the Boot Cotton Mills in 1849. Water enters through fixed guides (E) around the perimeter and then moves toward the center, through the vanes of the moving runner (D). The path is inward and downward, as shown by the arrows. Reprinted from James B. Francis, *Lowell Hydraulic Experiments* (Lowell, 1855), pl. 8.

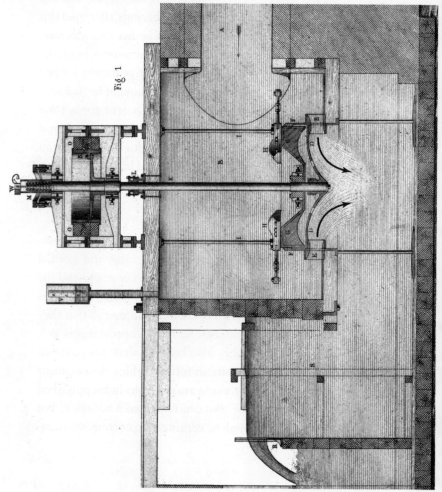

Fig. 1

sionally with Francis on technical subjects, the two men no longer worked in active collaboration. Boyden later said that by the 1860s, he had "chiefly relinquished designing turbines."[17] That must have been a disappointment to the directors and to Francis.

Boott Cotton Mills got Francis's centre-vent wheel, which incorporated the bucket form that both men claimed as their own. Francis has received more credit for this development, and wheels of the mixed-flow type are now called "Francis turbines" throughout the world. While the small Boott wheel of 1849 was a better prime mover than the simpler inward-flow wheel he had already installed at the Pawtucket Gatehouse, it did not match the performance of Boyden's best outward-flow turbines. Francis later published his somewhat disappointing but still promising results. His wheel had achieved a peak efficiency of 79.7 percent. We have no comparable test results for the "converging" turbine built on Boyden's plans for the Gay & Silver Company in 1854. Beginning in the late 1850s, other turbine designers, most notably Asa Swain, a former artisan in the Lowell Machine Shop, would make improved variations of the mixed-flow turbine into highly efficient and very popular prime movers. In many of the later runners with spoon-shaped buckets, water moved inward, downward, and outward.

The Lowell Machine Shop, working in association with Locks & Canals, became an important producer of Boyden outward-flow turbines in the late 1840s and continued to sell significant numbers of these wheels in and around Lowell. The *Lowell Courier* reported on August 30, 1870, that "most of the wheels of this city have been built at the Lowell Machine Shop; which establishment does an important business in this line."

Francis was responsible for much of this profitable activity. After the agreement with Boyden was in place, he prepared designs and supervised construction of four outward-flow turbines that were installed at the Tremont and Suffolk mills in 1851 and 1852. He was free to copy what he wanted from Boyden's previous turbines or to improve on them in any way he chose. One of his Tremont wheels became the subject of the most exhaustive turbine tests in *Lowell Hydraulic Experiments,* a book that he published in 1855. Francis said that Boyden provided only limited assistance as he completed his investigations of turbines for his book: "Mr. Boyden has communicated to him copies of many of his designs for turbines, together with the results of experiments upon a portion of them; he has communicated, however, but little theoretical information."[18]

Francis wanted Boyden to make his work available to other engineers. He wrote that Boyden had "accumulated a vast number of valuable experiments and observations upon them, which, it is to be hoped, he will find time to prepare for publication."[19] Unfortunately, Boyden was apparently too secretive to share his hard-earned knowledge and experimental data with a wide audience. He never published anything on turbine design or experimental procedures. Physical evidence of Boyden's turbine work was the basis for a chapter in *Lowell Hydraulic Experiments,* "Rules for Proportioning Turbines." Francis devised these rules by comparing particular Boyden turbines that had proven to be effective. Deciding on proportions was, however, only part of the challenge facing the inventor of a new turbine, and adhering to rigid rules could stifle creativity.

Boyden and Francis followed a set of broader principles that were apparently products of their reading in European hydraulic literature. When Francis was completing his centre-vent wheel for the Boott Mills in 1849, he wrote a letter to Boyden in which he listed three "considerations" that had most influenced his work. The first was a restatement of Lazare Carnot's fundamental dictum of turbine design: that "the water should enter the wheel without shock and leave it with the least practical velocity." The second consideration dealt with avoiding sudden changes in velocity or direction as the water went through the wheel. The third said that all the force of the entering water would be "imparted to the wheel" except for frictional losses and any velocity remaining as the water left.[20] Francis and Boyden were both trying to make water move in a smooth curve without rapid shifts in direction. By avoiding turbulence and by causing the water to exit with little or no absolute velocity, they sought to transfer as much energy as possible to the moving runner. These worthy goals applied to turbines of any type. In understanding how a turbine operated, Francis clearly benefited from Boyden's insights and criticism, which caused him to modify the original wording of his second consideration.

Both of these talented designers had more faith in experiment and graphical methods of analysis than they had in purely theoretical deduction. One of their basic techniques was to calculate "the path described by a particle of water in passing through the wheel."[21] Boyden had developed this method during his modifications of the Fourneyron turbine. Francis made use of the approach when he worked on one of his first designs for an inward flow turbine in January 1847. A more rigorous application appears in *Lowell Hydraulic Experiments*, complete with an illustration of paths for the outward-flow Tremont turbine.

We have already seen that the formula for calculating flow over a weir was a crucial part of turbine testing. Francis was critical of the procedures that European hydraulicians had used to study weirs, and he did want to see if their formulas could be improved. The basic weir formula was grounded in the seventeenth-century efflux theorem of Torricelli, which Francis called "the fundamental law of hydraulics."[22] But engineers had learned that you could not ignore the significant effects of factors such as friction and contraction. Theoretical formulas had to be tested and corrected before they were useful in engineering practice. In 1878, Francis told an Italian professor named Luigi D'Auria that "I fear I have a habit of doubting purely theoretical deduction in an unreasonable degree." He was clearly upset by that scholar's failure to test new ideas before publishing them. In the draft copy of his letter, Francis said that, "notwithstanding the great attention that has been given to the subject during the last two centuries by many men of distinguished abilities, those of your country taking decidedly the lead, very little of what we call the science of hydraulics has any practical value except that founded directly on experiment." Francis was always a polite and hospitable man, but he crossed out what was going to be an invitation for D' Auria to visit Lowell.[23]

His distrust of theoretical hydraulics was echoed by a Harvard-educated, German-trained engineer named Clemens Herschel, with whom Francis worked on engineering experiments after the Civil War. Herschel believed that theoretical hydraulics had "proved a wholly barren field for several centuries."[24] Both men used theory when it could help them predict the behavior of water or guide the direction of experiments, but they sought empirically determined correction factors that would bring theoretical predictions closer to reality. For them theory was only a starting point.

Storrow, who had studied with theoreticians in Europe, was more generous in his assessment of their contributions. He saw "beautiful examples of the application of mathematical analysis to the discovery of physical truths" in hydraulics, but he also supported the need for testing. "It is true," he wrote, "that many experiments are yet to be desired in order to confirm some of the results which theory alone has announced."[25] When Sir Cyril Hinshelwood looked back at his field later in the century, he observed that "fluid dynamicists were divided into hydraulic engineers who observed what could not be explained, and mathematicians who explained things that could not be observed."[26]

Francis was critical not only of the Europeans' emphasis on idealized mathematical theory but also of their tendency to rely both on models of less than full size and on testing equipment that could not handle heavy loads or substantial flow rates. Experimenters in Britain and the continent often cut corners in this way: "Owing to the difficulty and expense of making experiments on the flow of water on a scale comparable to that found in practice when water is used as a source of power, most of the experiments have been made with apparatus so small that the results obtained are not a good guide for the engineer."[27]

Because of his position as both agent and chief engineer of a great waterpower company, Francis could get the technical staff and the financing to perform large-scale hydraulic tests. The textile corporations that had taken over Locks & Canals had much to gain from the accurate measurement of water flow and turbine performance. Francis found that corporate officers usually understood the value of precise hydraulic information and were always "willing to defray such expenses as were necessary, in order that the experiments might be made in a satisfactory manner."[28]

A better weir formula would not only improve turbine testing; it would also allow Francis to develop practical ways to measure water usage throughout the recently expanded canal system. He could then monitor flow to the mills and halt any abuses in dry seasons. He knew the new dam at Lowell and the northern lakes that had been converted to storage reservoirs would provide much more than the flow promised in the existing waterpower leases. Once they had a realistic estimate of available water and industrial demand, the directors of Locks & Canals could increase the total number of millpowers to be leased and negotiate their redistribution among the manufacturing corporations. The canal company could even sell surplus water when it was available. But to control and bill for all this water, Francis first had to gauge it. Weirs were the tools he needed to begin this work.

Francis gave Boyden credit for suggesting the framework of the weir formula in 1846. This was a stripped-down formula without any numerical values for coefficients or exponents: the actual numbers were to be found through experiments. Francis, as we have seen, was already aware of European formulas, including those of Poncelet and Lesbros. The experimental work of these two French engineers was done at a small scale, but its precision impressed Francis and was helpful in the development and refinement of his formula.

The key exponent and one coefficient of the basic weir formula could be calculated relatively simply by running the same water through two weirs of similar form but different length and then comparing data. Francis did that at the Tremont mills in 1851 by modifying some of his turbine-testing apparatus. Much more ambitious experiments were necessary to perfect the weir formula. Find-

A weir experiment in progress at the Lower Locks. At the right is a pivoting gate, which Francis designed to open or divert water flowing over a weir. In the center a technician uses a Boyden hook gauge to measure the elevation of water approaching the weir. Another technician in the small building monitors an electrical signal, timing intervals of flow with a marine chronometer. The lock chamber at the lower right serves as the measuring vessel. Reprinted from James B. Francis, *Lowell Hydraulic Experiments* (Lowell, 1855), pl. 13.

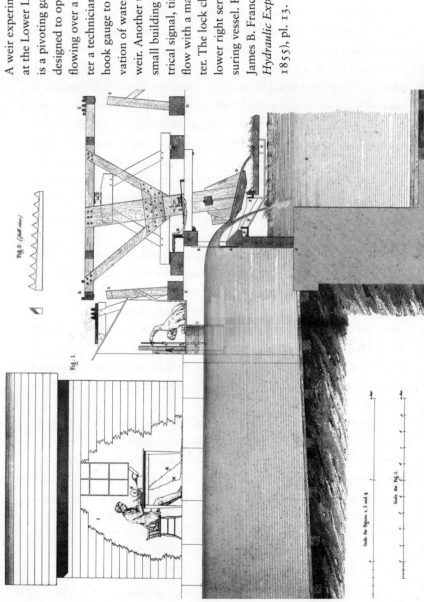

ing the principal coefficient was the most difficult challenge. Francis had to know exactly how much water went over a weir in a given period of time, and he intended to determine this with the sizeable flows used in Lowell. The only way to do that was to trap and precisely measure a great volume of water.

Francis decided to use one of the large, wood-lined lock chambers at the Lower Locks of the Pawtucket Canal as his measuring vessel. In 1852, workmen radically altered the lock for one series of tests. Across its entrance, they placed a temporary weir. They also installed a huge, weighted gate, which pivoted rapidly to open or cut off flow from the weir to the lock chamber. An electrical signaling device connected to the gate alerted a technician, who timed intervals of flow with a marine chronometer. Other assistants used Boyden hook gauges to record elevations of water to the thousandth of a foot. Francis even filled the indentations of nail holes and factored in the slight leakage from the tightly caulked chamber during every trial. His careful recording of experimental procedures, observed data, and calculated results met the highest standards of scientific inquiry. He was using scientific methods in pursuit of engineering goals. Experimentation had always been a part of American technology, but no one had applied this degree of rigor and precision in engineering practice.

In modern hydraulic manuals, there are many formulas for flow over sharp-edged weirs, but the one developed by Francis is always prominent in the list. The range of choices shows how variations in current of approach, length of weir, or other conditions can affect flow measurements. Francis simplified his formula deliberately and specified the situations for which it was appropriate. For instance, he kept a theoretical exponent of 3/2 in the formula despite experimental results that found 1.47 to be a more accurate figure. He explained that 3/2 was close enough for engineering purposes and was simpler for "persons for not well skilled in the use of logarithms."[29] The application of theory by engineers often involves making reasonable approximations and giving more weight to some factors than to others. Clearly stated restrictions also play an important

THE FRANCIS WEIR FORMULA

The following diagram explains how to apply James Francis's weir formula to determine the flow of water in cubic feet per second (cfs). A weir is an open aperture over which water spills. In his experiments, Francis used a rectangular weir with a sharp-edged sill. He measured both the length of the weir and the head (height) of water above the sill. He also allowed for the effects of end contractions, when the projecting sides of some weirs narrowed the sheet of water.

The Francis weir formula is $Q = 3.33 (L-0.1nH)H^{3/2}$

Q = Discharge in cubic feet per second (cfs)
L = Length of weir
n = Number of end contractions
H = head on weir (height of water above the sill, measured just upstream of the weir)

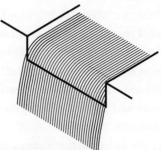

This view shows a weir with no end contraction, so n = 0 in this case.
With no end contraction, the formula becomes $Q = 3.33 LH^{3/2}$

The bottom view shows a weir with two end contractions, so n = 2 in this case.
With two end contractions, the formula becomes $Q = 3.33 (L-0.2H)H^{3/2}$

role. Francis's formula remained popular with twentieth-century engineers because they appreciated its ease of use and understood its limitations.

In 1855, Francis published the results of these and other experiments in *Lowell Hydraulic Experiments*. That handsome volume included tests of both the Boyden outward-flow turbine and his own inward-flow turbine. The meticulous study, which he went on to revise and expand in three later editions, was the first American book on engineering science to win international acclaim. It was largely because of the enthusiastic reception of his book in Europe that mixed-flow turbines became known as "Francis turbines." Francis did not continue his work as a turbine designer after the book's initial publication, and he made few direct contributions to the further development of mixed-flow wheels.

Anyone reading his book and studying its superb illustrations (engraved in France) could see the great care that Francis put into the fabrication of testing equipment. Engineers were equally impressed by his concern for precise measurements and accurate tabulation of data. As William Worthen observed in 1894, *Lowell Hydraulic Experiments* is "not only standard as to the rules therein given, but will always be standard as to the manner such experiments should be carried out, with appliances beautifully designed and adjusted, observations carefully taken and checked, and results elaborately but simply recorded."[30] In the same year a trio of distinguished engineers (including Worthen) concluded that "this work forms an era in hydraulic engineering." They were careful to note that the experiments "were not undertaken as a purely elementary problem in hydraulic science." Instead, they "grew out of the necessity for a practicable determination of the amount of water distributed to the various mills in their daily operation under every possible contingency of flow."[31] Mansfield Merriman, whose influential *Treatise on Hydraulics* went through ten editions by 1916, said that "the work of Francis on the experiments conducted by him at Lowell, Mass., will always be a classic in American hydraulic literature, for the methods therein developed for measuring the theoretic power of a waterfall and the effective power utilized by the wheel are models of careful and precise experimentation."[32]

Perhaps the greatest contribution of Francis and Boyden was their demonstration that systematic, verifiable test procedures were important in the process of engineering design. The turbine builders who came after them did not immediately adopt the scientific rigor of their experimental methods and their graphic analysis, but performance testing did become an essential part of wheel development.

John R. Freeman, an experienced hydraulic engineer with a degree from the Massachusetts Institute of Technology who knew and admired Francis, pointed out the difficulties of computing the path of water in designing turbine runners. It worked well in the early turbines of Boyden and Francis, "but after this path had been twisted in three dimensions by inward, downward, and outward discharge, it became impossible of description in equation or diagram." The turbine designers who soon followed these pioneering figures were less analytical in their methods. Solutions were "found chiefly by scale models in the hands of men with sound mechanical instinct in leading water without eddy round gentle curves and who then tested their new models in the laboratory otherwise called the testing flume."[33]

The word *flume* can be broadly defined as "an artificial channel for a stream of water to be applied to some industrial use."[34] In the nineteenth century, there were inclined timber flumes, through which lumberjacks sent their cut logs, flumes for placer mining, flumes that directed water to mill wheels, and flumes built in canals for measuring flow. The facilities for evaluating turbines became known as testing flumes. Over the course of numerous experiments, Boyden and Francis worked out effective layouts for such flumes and found or designed equipment that could measure time, head, flow rate, gate settings, rotational speed, and power. At the Pawtucket Gatehouse, which sometimes served as a testing flume, you can still see remains of the weir that Francis installed in the tailrace of his first operational turbine in 1847.

In what has been called the "cut and try" period of turbine design, some wheel manufacturers installed their own flumes to help them improve their designs, but most came to rely on commercial facilities for final evaluation of their wheels. Savvy customers soon insisted on seeing certified results from a reputable testing flume before they would buy a turbine.

Commercial testing flumes got their start in Lowell, with significant input from Francis. In 1869, Asa Swain decided to build a flume to test his own turbines. He hired James Emerson to supervise the construction of a flume based on designs by Francis. Emerson, a former sea captain with impressive technical aptitude, had already built dynamometers (like the Prony brake) for several turbine makers, including Swain. The chosen site for the flume was at a spillway on the Wamesit Canal, which used water from the Concord River and was not part of the Locks & Canals system. The object was to create a facility that could be used over and over with a variety of wheels. Until this time, Francis had done his full-scale tests at the sites where the turbines were installed. The labor-intensive nature of Francis's experimental procedures and the necessary installation of measuring apparatus, much of it custom built or adapted to fit the particular situation, made the costs of testing very high.

A Lowell newspaper explained why a less expensive system of testing would be a benefit for turbine purchasers. The "very elaborate experiments" that Francis had done "were probably more reliable than anything of the kind ever previously made: but they were very expensive, so much so as to prevent wheel manufacturers of limited means from following them; and the sale of wheels has depended more upon the glib tongues of venders than the actual merits of wheels."[35]

Another major benefit of extensive turbine testing was the compilation of data that would allow turbines to be used as water meters. Tests that included measurement of water discharged by a turbine at various speed-gate settings and heads were particularly valuable for waterpower companies that subdivided available flow among multiple users. Speed gates controlled the flow onto the runners of turbines. By regularly checking the speed-gate settings and the effective head at mills that owned previously tested turbines, a company like Locks & Canals could get a good sense of how much water was going through those wheels. Although not perfectly reliable, for reasons to be explained, this was one of the easiest ways to regulate water use and charge for flow actually used.

In managing the Lowell Canal System, Francis had been monitoring turbine speed-gate settings in some mills since at least the early 1850s. His 1849 tests of the wheel that he designed for the Boott Cotton Mills were still being used to help determine discharge at that mill in 1868. At the Lawrence Manufacturing Company, Francis "deduced" the discharge of an 1854 turbine from earlier tests of a similar wheel at the Merrimack Manufacturing Company.[36] In that case, it was necessary to factor in the differences between the two wheels and the heads on which they operated. This calculated data was probably for emergency use only, since Francis had other ways to measure water consumption in the Lawrence mills. In calculating discharge, he much preferred to have actual test results for particular wheels at a wide range of gate openings. By promoting testing flumes like the one set up by Swain, Francis hoped to gain not only better discharge data but also other important information for assessing turbines.

Swain and Emerson made the flume at the Wamesit Canal operational, but, as Emerson put it, "the company was urged to employ an engineer with at least a theoretical knowledge of such tests." It seems likely that Francis made this suggestion, because Swain asked one of Francis's former assistant engineers, Hiram Mills, to run the formal tests with Emerson's help. During one evaluation, both Francis and Mills provided direct supervision when the managers of the Massachusetts Cotton Mills in Lowell were considering the purchase of a Swain turbine. Swain also paid for one series of public tests in 1869, open to any makers who wanted to submit a turbine. Robert Thurston said that this competition was the beginning of "a general system of public tests of turbines at testing flumes open to all users and builders of wheels."[37]

Turbine testing was grueling, labor-intensive work. The technicians who did this had to take precise measurements with gauges, operate machinery, read stopwatches accurately, and stay alert for any anomaly that might render a trial invalid. Francis had adopted Boyden's practice of using several people to make observations of the same gauges, seeking greater accuracy through redundancy and comparative readings. Repetition of tests and deliberate variation of parameters such as turbine gate settings or head extended the process over several days or weeks in some of Francis's experiments. Careful recording of observations and laborious calculation were essential. The key was to avoid mistakes in carrying out the tests and in computing the results. Competent, well-trained employees were necessary at both the test site and the office.

Emerson had worked enough with Francis and Mills to appreciate their concern for accuracy, and he maintained high standards at the Lowell flume when he took over its operation in 1870. He apparently bought the flume and continued to pay the Wamesit Power Company for the water he needed. In running the facility as a commercial enterprise that tested turbines from any maker for a fee, Emerson achieved efficiencies of operation that made the flume's services affordable for a wide range of turbine builders.

According to a local reporter, Emerson was "a practical believer in 'woman's rights.'" At a public demonstration of his flume, "he employed young ladies to observe the gauges." Emerson argued that the women were "accurate, attentive, and more reliable than boys." One of his regular helpers was a close relation: "His daughter, through the kindness of Mr. Francis, having obtained some insight into the method of measuring water, assists at tests, and renders valuable assistance in making the computations."[38]

The Holyoke Water-power Company invited Emerson to set up a flume on the Connecticut River, with water provided at no charge. Emerson's subsequent tests at the facility he set up in an unused Holyoke lock were very important in both the development and the marketing of turbines from a wide range of builders. Despite Emerson's abilities, his abrasive personality and lack of professional standing continued to cause problems.

Like Boyden, Emerson found fault with many people who built, sold, or tested turbines. The two men were, however, very different in their dealings with the public. While Boyden was naturally secretive and did not publish his technical procedures, Emerson became a widely read author with his *Treatise Relative to the Testing of Water-Wheels*. This strange, idiosyncratic volume, containing test results as well as personal philosophies and cartoons mocking "The Law" and the "Massachusetts School System," went through multiple editions and focused attention on the Holyoke testing flume.

Emerson resented engineers with academic training or with acquired skills in higher mathematics and hydraulic theory. At a time when the fields of civil and mechanical engineering were becoming professionalized, he was not considered a fully qualified engineer. The Holyoke Water-power Company eventually took over his flume and put well-educated men like Clemens Herschel in charge of tests. In 1882, Herschel designed a new flume in a specially constructed building between two canal levels.

Thurston, the first president of the American Society of Mechanical Engineers, gave qualified praise to Emerson's turbine reports in an 1886 article but left no doubt about his status: "The reports were found, by hydraulic engineers, to be full of valuable data, and, although not systematically arranged, or so complete in the analysis of the distribution of useful and lost work as a professional might have made them, form an extensive and valuable collection of figures." Francis's reports, on the other hand, embodied "the most painstaking analysis." Thurston added that the "later work of the Holyoke Water-power Company, which combines the facts and data with careful and skillful analysis, is likely to prove still more valuable."[39]

The methods of turbine testing in Holyoke were similar to those developed by Francis, but Herschel added his own refinements to the equipment and to many of the recording and calculating procedures. Thurston, after observing Herschel's tests in 1883, noted that "the calculation of flow [was] by Francis's formula, as is customary throughout the U.S., wherever weir measurements are used."[40]

Francis continued to do occasional tests of newly installed turbines, publishing articles on wheels by Swain and by Humphrey. Although previous turbines had often been custom made at great cost, American makers like Swain and Leffel soon put out catalogs of standard models in multiple sizes, suitable for a wide range of heads and flow rates. Higher operating speeds and better ratios of size to power, combined with economies of mass production, made this technology a bargain by the mid 1870s. Still, buyers wanted authoritative test results to help them make purchasing decisions. The demands of waterpower users helped establish the practice of testing industrial equipment and consumer products.

In his many hydraulic experiments, as in his structural investigations of beams and columns and his studies of boilers and smokestacks, Francis was seeking guidelines for other engineers. He was more concerned with reliable methods and useful formulas than with theoretical explanations for phenomena. As a corporate employee, he may have felt that he had to justify his more scientific work as a benefit for those he served. When, on occasion, he did investigate an underlying principle in hydraulics or mechanics, Francis no doubt expected that deeper scientific understanding would soon lead to useful advances in technology. Francis argued that his experiments were done "in the discharge of his duty, as the Engineer of the Corporations at Lowell."[41]

Francis purchased most of the important publications on hydraulics and mechanics that were available. He also had things copied when book dealers could not supply originals. He warned that "unless an engineer buys all the newest books & transactions of many learned societies, not only in English but in French, German & Italian, or has access to them in some way, he cannot be sure he is fully up to the times." That was particularly difficult to do in America, where he thought all engineers were "behind in different degrees."[42] Over time, he made the Locks & Canals library a great resource of technological information.

The chief engineer was expected not only to provide technical information but also to settle the inevitable conflicts that arose among the textile corporations on his system. Worthen saw Francis's methods as a model for "preserving amiable relations and cutting off interminable and expensive law suits." At many waterpower sites in the nineteenth century, litigation was a costly intrusion on normal operations and a serious distraction for industrial managers. Procedures for arbitration of serious disputes were in place at Lowell after 1848, but companies usually trusted the judgment of the chief engineer. He tried to improve the efficiency of the entire system, to increase the total amount of power available, and to see that each corporation took only its share. In general, he got good cooperation from the textile producers, because instabilities in this dynamic system were disruptive and thus counterproductive. There was relatively little competition between these firms, which often shared directors. Francis was there to make things run smoothly. Worthen accurately titled him the "chief of police of water."[43] What he needed most for the essential policing of the system was accurate and timely information about the amount of water that each corporation was using.

Francis knew that he could not install a permanent weir in a power canal or headrace to check flow without giving up valuable head. Setting one in a tailrace might also reduce net head. The chief engineer needed a practical, unobstructive way to make frequent measurements of the water used by each mill complex. He had tried a surface float, but, even with empirically determined correction coefficients, it did not give an accurate enough measurement of flow. Now he turned to another device that promised much better results: a floating rod or tube.

Thomas Mann had argued in the Royal Society's *Transactions* in 1792 that one could measure the velocity of moving water with a wooden rod. In 1852, Francis tried a variation on Mann's method. The chief engineer timed the transit of a floating tube through a wood-lined flume built in the canal leading to the Middlesex Manufacturing Company. Using a hollow metal tube weighted with lead so that it would remain vertical, with its lower end just above the bottom, his men recorded the time of transit between two beams and then calculated the velocity of the tube. He assumed that a tube would travel with approximately the same speed as the water that moved it. What he really wanted was the mean velocity of water through the flume. As he knew from his work with surface floats, the mean velocity (feet per second), multiplied by the cross-sectional area of the flume (square feet) would give him flow (cubic feet per second) and tell him how much water a corporation was using.

In Francis's experiments, the same water that went through the flume passed over a temporary weir, where flow was measured again. In that way, he could directly compare the results of weir and tube measurements. His initial experiments with tubes were tentative and did not involve as complete a distribution of tubes across the entire cross section as he would try later. In the first edition of *Lowell Hydraulic Experiments,* the flume experiments are the weakest contribution, but the early results were still very promising, and he intended to do more investigations.

In 1855 and 1856, Francis completed a much more thorough series of experiments with flumes and floating tubes. His men recorded not just the time of passage but the horizontal points where the tube crossed each transit beam. Repeated trials, with the tube inserted approximately every foot and a half across the width of the flume, provided enough data to plot a graph of tube velocity versus distance from one side. Francis realized that the actual velocity of the water varied across the flume and from the surface to the bottom and that the tubes did not record some of that variation. Using weir measurements for comparisons, he devised a simple correction formula that took into account the layers of slow-moving water at the bottom and sides, which his tubes missed.

In time, boundary-layer theory would explain why water slowed near the perimeter of a flume. Francis had no immediate need for this or other sophisticated theories of flow. His correction took care of multiple uncertainties in determining the mean velocity of the water from the observed movement of the tubes. He did acknowledge that there were ways to estimate the mean velocity using complex mathematical calculations and theories of both tube motion and velocity distribution. "It would, however, involve lengthy computations, and the result would not be free from uncertainty . . . and however interesting such an investigation might be as a scientific matter, it will be safer in practice, to rely upon rules deduced from suitable experiments, even if such rules are empirical."[44]

Francis used both empirical data and theory to show that tubes were a practical way to measure flow. One theoretical question was whether a tube actually floated at the same velocity as the mean of all the currents of water that acted on its immersed length. Joseph Frizell, a gifted mathematician, helped his chief engineer explain what happened to a floating tube. Francis was not averse to theoretical arguments and knew their power to persuade. The complex explanation of tube velocity in later editions of *Lowell Hydraulic Experiments* leads to the conclusion that differential forces act to both retard and accelerate the tube as it moves with the current but that these forces generally cancel each other out. Long after Frizell had left Locks & Canals, he included discussion of this issue in his own book, *Water-power*. He credited Francis for his "mathematical investigation" of the retarding effect of water pressure. He also mentioned "the labors of physicists" who determined the accelerating effect of the inclination of the water surface. Agreeing with Francis, he said that the effects "would practically offset each other" in the typical current of a flume. It was safe to assume that the tube moved with the mean velocity of the water that touched it.[45]

By the late 1850s, Francis had thoroughly tested the use of floating tubes and had installed six permanent flumes in the canals. As we see in chapter 5, he came to depend heavily on tubes for management of his waterpower system. They gave very accurate results, and they measured all the water drawn by a mill complex, including the large amount of process water used at some sites and any leakage through or around turbines. Relying on turbine gate settings to indicate flow, as he did at some mills regularly and at others only in special situations, was not as inclusive or as reliable. Even when he had good test data on a particular turbine model, he could not be sure that it remained in perfect condition or that its speed-gate settings were correct. Damaged or broken gate registers were not unusual. Debris that lodged in the passages of guides or runners was another source of error. The elaborate tables of discharge that he produced were no help whenever a mill changed turbines.

The expanded 1868 edition of Francis's book provides detailed descriptions of his conclusions about tubes. In 1878, he wrote an article for *Engineering News* in which he reported on a demonstration for members of the American Society of Civil Engineers who made a trip to Lowell that year as part of an annual conference held at the Massachusetts Institute of Technology (MIT). The figure on p. 138 is a photograph taken at that meeting. The man standing with a tube in the lower left corner is a key member of the "permanent corps of Hydraulic Engineers and Assistants," who did "a great amount of gauging" for the Proprietors of Locks and Canals.[46] He is about to put his tube into the swift water of the Merrimack Canal. Observers are seated on the first of two transit beams.

Flume measurements were carefully choreographed operations. An engineer or "trained assistant" was the recorder in charge of a crew of three men. Frizell described him starting at the upstream transit beam, "stop watch and notebook in hand, pencil in mouth; the notebook ruled with appropriate columns." Another important figure was the "intelligent workman" who launched the tube from an upstream bridge with a "movement requiring some skill."[47] The recorder started his stopwatch when the floating tube crossed the face of the first transit beam. Then, after walking quickly to the second transit beam, he noted the elapsed time of transit. Meanwhile, one of his men was observing the distance of the tube from one wall of the flume as it floated by each transit beam. The company painted numbers, indicating distances in feet, on the beams, starting at the left side and ending at the right side. The final crew member leaned from a downstream bridge and retrieved the tube from the water. To plot an effective curve of velocities required many trials, spread as evenly as possible across the flume. Data from Locks & Canals records, as well as the author's imitative experiments with a small-scale model of a Lowell flume, suggest that the

A demonstration of flow measurement for the American Society of Civil Engineers in Lowell in 1878. At the lower left is a Locks & Canals employee with a long tube, which he will float down the Merrimack Canal, through the wood-lined section that served as a testing flume. Observers sit on the first of two transit beams, which are exactly seventy feet apart and marked to indicate feet of distance from one end. Courtesy of the Center for Lowell History, University of Massachusetts Lowell.

tubes tended to remain on course, moving roughly parallel to the sides.

Despite his preference for the floating tubes, Francis stayed well-informed about alternative procedures and used other methods when expedient: "The mode of ascertaining the quantity of water drawn was left open, so that the method best adopted to the circumstances in each case could be adopted, and also all of the improvements which might be suggested, from time to time, by the progress of the science of hydraulics."[48]

Current meters were one of the options available to Francis and favored by many engineers for measuring flow through a channel. These mechanical instruments, which had a propeller or rotor turned by the current, could be lowered from the same bridges used to launch tubes in Lowell's flumes. Manufacturers or private testing facilities calibrated, or "rated," individual meters so that revolutions per minute of the rotor could be converted to feet per second of current. The common method used in canals was to take spot readings at multiple points in a cross section (ideally, within a well-constructed flume).

Over the years, Locks & Canals bought a number of current meters of the propeller type and gave them fair trials. They may have been useful in estimating flow in the Merrimack River, but Francis found that they were not as reliable, as efficient, or as accurate as the tubes for determining flow in the canals. In the 1868 edition of *Lowell Hydraulic Experiments*, he said that a current meter depended on timed measurements taken with "delicacy" at a great many precisely located points in the cross section of a flume. The chief engineer explained that this would take too much time "in our large scale channels."[49]

He may also have been frustrated by the fragility of the meters, which had to withstand rough handling in the field and were adversely affected by sediment, weeds, and leaves drifting with the current. Anything that increased friction on the rotor shaft or interfered with the rotor vanes would affect meter performance. A damaged shaft could render a meter inoperable, as Francis learned when he had to send one out for repairs in 1881.

Cold weather in New England affected current meters. Members of a measuring team had to break up or cut a series of holes in surface ice before they could use a current meter. Even then most meters were unreliable if there were drifting ice particles (anchor ice) in the stream. Tube floats would not work if stationary surface ice covered any part of the flume, but they were unaffected by small pieces that moved with the current.

In 1879, Herschel, then chief engineer of the three-level water-

Flume and Float Measurements

It took considerable skill to insert a hollow tube of sheet metal into the swift current of a power canal from a footbridge. Once released, the tube floated downstream through a wooden flume at almost exactly the same velocity as the water carrying it. It was carefully weighted with lead so that it would remain vertical, with its lower end just above the floor of the flume. An employee with a stopwatch recorded the time a tube took to cover the distance between two transit beams (marked a and b). The distance divided by those times would give tube velocities. Another man recorded the positions where the floating tube crossed those beams. Repeated trials, with the tube inserted approximately every 1.5 feet across the width of the flume provided enough data to plot a graph of velocity versus distance from the side of the flume. The images here recreate an actual measurement at the Boott flume on May 17, 1860, but not all of the team members are shown.

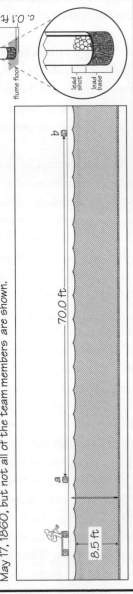

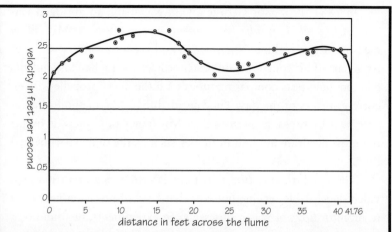

The plotted points on the graph represent the velocities and mean positions recorded by Lowis Jackson on May 17, 1860, and published in his book *Hydraulic Manual* (London, 1883), 142-47. The total area beneath the curve of these points, when divided by the width (representing the width of the flume), would give the mean velocity of the tube through the flume (2.431 feet per second in this example). For greater accuracy, there were mathematical corrections to allow for the slow-moving water at the bottom and sides of a flume that a tube did not reach. Flow (Q) through a flume was equal to the mean velocity (V) times the cross sectional area (A) of water in the flume: Q = VA. Q in this example was 863.59 cubic feet per second.

power system at Holyoke wanted Francis to conduct experiments with a current meter, but his friend was skeptical. A letter from Francis to an assistant makes that clear: "As to the current meter experiments, I think the chance of their being of much use in gauging the water used by the mills, is too remote to warrant hiring help specially for the purpose."[50] Francis's negative opinion of his current meter's value for regular gaugings did not change. He noted on February 22, 1881, that "the instrument has not been used for some time" and that its spindle was bent. Four years later, in the inventory taken at his retirement, the repaired meter was listed under "Contents of Closet . . . up Stairs" and not under "Water Measuring Apparatus" with the tube floats.[51]

In 1885, Professor George Swain was using an abridged version of *Lowell Hydraulic Experiments* as a textbook at MIT. Flow measurements, turbine experiments, and visits to waterpower sites were part of his course. As one class project, Swain had his students conduct field tests comparing tubes with the finest propeller-driven current meters of the day. They decided that "the method of measurement by tubes, as perfected by Mr. James B. Francis, is by far the most accurate method in flumes such as the one experimented on."[52]

There is little question that Francis's tube measurements provided dependable results, but was that level of precision necessary to monitor flow, or was it obsessive? In experiments, he found that his meticulous procedures (with application of his correction formula) gave flow figures that were rarely off the mark by more than 2 percent. The mean error was less than 1 percent. Inserting tubes at much wider intervals, or only in the center of the flumes, would have given less accurate but still useful results. A few spot readings with a propeller-type current meter would have been enough for most engineers, but not for the men who managed the Lowell Canal System.

Francis's precise, full-scale experiments, which were unprecedented in the United States, had lasting impact not only on the field of hydraulics but also on American research and development in general. What he was doing in Lowell with turbines, weirs, flumes, and tubes was scientific engineering. Francis always identified himself as an engineer, never as a scientist. He used theory, mathematics, and scientific recording procedures, but he was not seeking a theoretical understanding of hydrodynamics. He clearly distrusted the direct application of untested mathematical theories to solve real problems. His goals were practical and in line with engineering traditions: Francis wanted to *work* with water—to manage it, measure it, and predict its behavior.

CHAPTER FIVE

Protecting the People and the Profits, 1847–1865

Opening the Northern Canal in 1847 and controlling its flow with gates operated by an innovative turbine were tremendous engineering accomplishments. Francis, however, was not finished with additions and modifications of the system. Taking maximum advantage of the new canal and the reservoirs in New Hampshire would require more changes, and the chief engineer was growing increasingly concerned about both natural and man-made hazards. His highest priorities were to protect the canal system, the mills, and the people of Lowell. His role as the principal engineering consultant for the corporations expanded, as did his involvement in the political affairs and physical security of the city. It would not be easy to manage Locks & Canals, to distribute water fairly, and to reduce or eliminate damage from floods and fires.

Francis's contributions to turbine design and other aspects of scientific engineering have given him a permanent place in the international history of technology, but a very simple gate has enshrined him in the pantheon of Lowell heroes. The chief engineer is considered the savior of the city because he recognized Lowell's vulnerability to the terrible floods that swept down the Merrimack River once or twice a century. In building a protective gate, however, he was also correcting a serious oversight: his Northern Canal project of 1846–47 had made an existing threat even worse.

A guard lock is one way that engineers shut off the flow of water into a canal. If its upstream miter gates have enough strength and height, and if walls and surrounding land are above flood levels, the lock should be able to hold back the torrent. The Guard Locks site in Lowell included a special dam or embankment, but it was also necessary to close the separate set of sluice, or head, gates that regulated normal flow to the mills. Flood waters surging through the Pawtucket Canal into the heart of the city would cause tremendous property damage and might take lives as well. The entire Guard Locks complex was the key line of defense.

The main problem with flood protection in nineteenth-century Lowell was that people kept building and altering structures at Pawtucket Falls. Dams and bridge piers in the channel became serious liabilities during major freshets. Water would rise as it tried to go over or around them. In 1831, a year after Locks & Canals rebuilt its Pawtucket Dam on a new alignment, a flood nearly got past the guard lock on the Pawtucket Canal. Arthur Gilman recalled employees "throwing up a temporary embankment of earth" at the north end of the lock. For several anxious hours, he feared that the flood would prove "disastrous to a large portion of the town."[1]

Although everyone was relieved when the water receded, the managers of Locks & Canals recognized that the city would not be safe without changes at the Guard Locks. In 1832 and 1833, superintendent Moses Shattuck constructed a higher guard dam of stone and replaced the existing sluice gates (for controlling flow to the mills). The new height was apparently based on an estimate of the highest water level in 1785, which was the worst freshet anyone could remember. In 1833, however, Locks & Canals also added two feet of granite headers to its Pawtucket Dam (see p. 49).

After the new dam had been in place for several years, Locks & Canals hired one of its former staff to study the potential for flooding. Uriah Boyden took on this assignment long before he became a famous turbine designer. Boyden interviewed a number of elderly men who had distinct memories of the 1785 flood. He even found a nail that one of them had hammered into a tree to mark the height of the river on that frightening occasion. With this and other evidence, Boyden estimated the levels at various points from Nashua,

New Hampshire, to the dam at Lowell. He thought that water would have been 13 feet 5 inches above the 1833 dam.

Severe freshets seemed to be getting more frequent. Two days of rain in the winter of 1841 "raised the rivers, very suddenly, to an almost unprecedented extent." The *Lowell Courier* said that the flood level was higher than anyone had seen "for many years." Ice and debris in the water made this freshet particularly spectacular. "The Falls afforded a grand and beautiful sight . . . and were visited by hundreds of our citizens"[2]—a memorable scene, but also a disturbing one for those concerned with the city's safety. Neither that 1841 flood, nor another in 1843, reached the 13 foot 5 inch height that Boyden reported for the torrent of 1785. So far, the Guard Locks had held, but had Lowell seen the river at its worst?

Francis changed the equation in 1846–47 with the Northern Canal project. Adding a boat lock and a brick gatehouse at the entrance to the new canal was a significant alteration. When he made these improvements to the system, he realized that they might have unintended consequences. The destruction of his cofferdam in a freshet on June 1847 was a warning that worse floods were possible.

Once again, Francis examined the issue of flood heights, this time considering the effects of all the additional things in the channel. He soon realized that these new structures, along with the older bridge piers, would push water to an unprecedented elevation if flow in the river ever matched that of the 1785 flood. During that torrent there had been "no artificial obstruction to the free flow of water at the head of Pawtucket Falls" (the bridge was not built until 1794). Francis "concluded that a height considerably greater than the height of the water in the freshet of 1785 should be provided for." He made sure that the new guard features at the falls could withstand the effects of water nineteen feet above the dam.[3] Then he considered what a great flood could do at the Guard Locks of the old Pawtucket Canal.

The chief engineer knew that high water above the dam would also enter the Pawtucket Canal. That waterway ran directly from the millpond into the heart of the city. His arguments convinced the directors of Locks & Canals that the threat from the river had actually increased. The dams and spillway controls of the Lake Company in New Hampshire might help, but they would not be enough to prevent great floods. On September 21, 1847, his directors authorized him to "raise the work at the Guard Locks on Pawtucket Canal so as to be safe against accident from extraordinary freshets."[4]

Francis wanted an impenetrable barrier across the canal that led into the city. The most difficult step was to raise the existing guard dam and extend it further on either side of the lock chamber and sluice gates. When he had finished, the top of this stone and earth dam at the Guard Locks was nineteen feet higher than the crest of the Pawtucket Dam.

Work on another critical component of this flood protection assemblage was not as rapid. In his 1847 estimate, Francis had included wood for a gate across the guard lock chamber on the Pawtucket Canal. He intended to keep the old miter gates at the head of the lock, but he was going to back them up with an enormous gate suspended above and spanning the width of the chamber. He designed this gate to function much like a portcullis at a castle: when dropped, it would deny entrance to the threatening water to the full height of the guard dam. During the alterations of 1848, he made vertical slots in the stonework of the lock and guard dam walls to guide the heavy gate as it fell and to hold it in place against the tremendous pressure of floodwater. Francis predicted that a great flood would some day top the miter gates. Still, he waited two years before erecting this "Great Gate," which could plug the gap at the lock.

His delay almost cost him dearly. In the winter of 1850, high water again caused trouble at the falls, but spared the city. Samuel Richardson, an official of the Pawtucket Bridge Corporation, thought that Locks & Canals should compensate his company for "damage done said bridge by the recent freshet." He claimed that the flooding was "in consequence of the new canal." Although the damage to the bridge was relatively minor, Francis was probably disturbed by the blame that Richardson placed on the Northern Canal project. This incident raised the specter of catastrophic liability for Locks & Canals in some future flood. The chief engineer said later that the piers and abutments of the bridge were partly responsible for "obstruction to the free flow," but he recognized that his gatehouse and lock at the end of the dam might have contributed to the problem.[5] Within months of Richardson's complaint, Francis finally installed the "Great Gate" for the Pawtucket Canal.

Carpenters assembled twenty-six white pine timbers with connecting components of iron to form the gate. Francis had subjected the wood to a Burnettizing process, which he had imported from England and tested in Lowell. This preservative treatment, using a solution of zinc chloride in an iron pressure chamber fabricated at the Lowell Machine Shop, was among the earliest American experiments in making wood resistant to decay and insect attack. Locks & Canals did a great deal of Burnettizing and Kyanizing (another wood-preservation process) for paying clients at its purpose-built plant, but it also treated many of the timbers, boards, and shingles used on structures throughout the canal system, where moisture could shorten the useful life of untreated wood.

The enormous gate, which still survives, is twenty-five feet high, twenty-seven feet wide, and seventeen inches thick. Because this gate weighs nearly twenty tons, its supporting structure had to be substantial. Francis's workmen hung the gate from a braced, heavy-timber frame with an iron shackle. Some of them probably believed that the gate would never be dropped. The general public was apparently even more skeptical. The precautions taken by 1850 were expensive, and the flood that Francis feared seemed little more than an apocalyptic dream to most of Lowell's citizenry.

Within two years, Francis was vindicated. The river demonstrated its capability to set another record for flood height. Four days of rain in mid April 1852 swelled the already heavy flows of spring. Before the water even reached the top of the upstream lock gates at the Guard Locks, it began to work its way around them. At 3:30 AM on Thursday, April 22, a workman from Locks & Canals frantically cut through the suspending shackle with a cold chisel.

Pawtucket Canal Guard Locks in 2008. The single lock chamber of the guard "locks" is at the left, with the Francis Gate hanging above it. At the right is the brick gatehouse where Francis used hydraulic rams to operate sluice gates for admitting water to the system.

Gravity did the rest. The vertical barrier dropped swiftly through its stone guides into the lock chamber, and an urban disaster was averted. Although the water continued to rise another two feet, the portcullis gate held firm. The derision that had been directed at "Francis's Folly" in 1850 turned quickly to acclaim. A Boston newspaper reported that "Lowell has probably escaped a great calamity by the wise fore sight and timely precaution of one eminently practical man—James B. Francis Esq."[6]

Although the flood of 1852 did not reach the flow rate of the one in 1785, water did rise fourteen feet and one inch above the dam, higher than the estimated crest of the earlier torrent and just above the miter gates, which had been the principal protection for the city until he installed his "Great Gate," in 1850. The chief engineer looked at the height that Boyden had established for the 1785 flood in Nashua, New Hampshire, and compared it with the recorded height at that upstream city in 1852. The water was two feet higher in Nashua in 1785. He concluded that less water had come down the Merrimack in 1852 but that "obstruction to the flow of the river at and near the head of Pawtucket Falls" made the river rise to an unprecedented level at Lowell.[7] The principal obstruction was, of course, the main dam, which had not been in place in 1785, but Francis's Pawtucket Gatehouse, lock, and guard dam made things worse. He was lucky that nature did not unleash its fury until his gate was ready.

After the gate saved the city, newspapers in both Boston and

The Francis Gate during the flood of 1936, when it again saved the city of Lowell. The gate is down, creating a barrier at the Guard Locks. Courtesy of the Lowell National Historical Park.

Lowell ran an article praising Francis's decision to "take measures against what *every one* considered a very remote possibility, by what many considered a useless expenditure of money." The reporter noted that "it is certainly somewhat remarkable that a freshet occurring only once in a century, should happen within two years after special preparation had been made for it." Without what was later called the Francis Gate, "every vestige of the old guard gates would have been swept away, and a mighty and uncontrollable river would have swept through the heart of Lowell, destroying everything in its course."[8] Later, a group of prominent citizens paid for an ornate silver service, which they gave to Francis for his "prudence and foresight in providing a sufficient guard against the freshet of 1852."[9]

In a much less dramatic way, Francis surely saved many mills, and perhaps much of the city, from another terrible threat—that of fire. Textile mills burned with disturbing frequency, despite continuous efforts from the 1820s on to improve their construction, lighting, and firefighting equipment. Inadequate water supplies, incompetent or feuding fire companies, congestion, and widespread use of flammable building materials made entire sections of nineteenth-century cities firetraps. Despite the presence of a professional fire department, sixty-five acres of downtown Boston burned to the ground in the Great Fire of 1872, only a year after an even worse disaster in Chicago. Yet Lowell was spared the type of devastating fire that leveled major sections of cities all over America. And Lowell's textile corporations, after some early setbacks, had no catastrophic losses from fire while Francis was the agent of Locks & Canals. Waterpower, canals, and engineering know-how help to explain that impressive record of fire suppression in Lowell.

A journal from the Lowell Fire Department in 1845 shows how big a role the corporations played in firefighting both inside and outside the mill complexes. Chief Engineer Bancroft listed fourteen engines. Of these, only three were city property, and one of those was kept in a building owned by the Middlesex Manufacturing Company. Engine number 1, purchased by the city in 1829, had originally sat in an engine house on Locks & Canals land, provided rent-free. Kirk Boott, who chaired the first meeting of the Fire Wards, was just one of many corporate executives who took part in Lowell's fire protection. Eleven engines belonged to the corporations in 1845, and most of those engine companies were fully manned. Locks & Canals enlisted a company of fifty-seven men, which kept an "engine and all its apparatus . . . in good condition." The city report went on to describe that engine as "a powerful one . . . .[that] throws two streams of water at the same time if necessary. It is of great service to the Fire Department when fully manned."[10]

At first, textile corporations in Lowell used force pumps to raise canal water up to cisterns in the attics of their mills. Waterwheels were the normal power source, although in later years, steam engines also did this work. From the elevated cisterns, water under pressure was available at hose connections on each floor. The corporations also relied on their own fire engines, which trained employees could use to direct water from the canals onto burning buildings. Interconnected systems for entire complexes, with underground pipes linking mills and multiple force pumps, became common after 1830. If a fire could be extinguished quickly, the water stored in attic cisterns might suffice, but their capacity was modest. It was usually necessary to start the force pumps during serious fires. That could present difficulties if the mill was closed. Worse problems arose when the canal that supplied a mill happened to be drained for periodic cleaning or repair; then there could be no waterpower to run the pumps and no canal water for extinguishing the flames.

In the late 1840s, Locks & Canals tied the individual fire mains of all the corporations on the canal system together so that their force pumps would be interconnected. This was not an ideal solution, but it enabled the mills to support one another with pressurized water in most situations. It also set the stage for one of Francis's finest achievements: construction of a reservoir that served every mill yard and also helped defend neighborhoods outside the mill gates.

Francis decided that what the corporations needed was their own reservoir holding a substantial volume of water at high elevation. Once connected to the fire main system, it would be a reliable source of water with enough pressure to reach the roofs of tall industrial buildings. In particular, he wanted "a reservoir to supply the mills of Lowell with water, in the event of a fire occurring when the water was drawn out of the canals."[11] The city had been considering construction of a public water system with one or more reservoirs that could supply the needs of both the corporations and the community at large, but nothing was happening to implement that expensive plan. Francis favored immediate, private action to protect the mills. The directors of Locks & Canals agreed with their chief engineer, who was already advising many of their corporations on issues of fire protection. In 1849, they let him build a two-million-gallon reservoir on Lynde's Hill, in what was then the Belvidere section of Tewksbury and is now part of Lowell.

This reservoir, positioned much higher than the cupolas on any of the mills, supplied pressurized water to hydrants located in the mill yards and near blocks of corporate housing. It was also connected to standpipes and hydrants within the individual mill buildings, and over time, to the sprinklers that Francis encouraged mills to install. Within a few years of opening, the reservoir took on an additional role as a source of boiler feed water and process water, for which companies paid a reasonable fee. Francis wanted the corporations to use their own force pumps to replace anything taken from the reservoir, but most of the water sent up to Lynde's Hill was provided by the Lowell Machine Shop, which used one of its breast wheels and a special waterpower allotment to pump from the Pawtucket Canal. Locks & Canals compensated the shop for this critical service.

In 1850, all the corporations signed an agreement that formed the Lowell Manufacturers Mutual Fire Insurance Company, and they made Francis the agent and fire inspector of the newly chartered company. He experimented with and promoted perforated pipe sprinklers, better hoses and nozzles, and slow-burning or fire-resistant construction. Later he evaluated patented sprinkler heads, one of which quickly extinguished a blaze in a storehouse of the

Middlesex Manufacturing Company in 1866. Francis set and enforced rigorous standards that reduced fire damage. His analytical approach to hazard assessment made him a leading authority on fire prevention and containment and a valued advisor to a number of communities, including Lowell. The corporations had always worked with the city to halt fires on both corporate and private properties. Any fire could spread quickly, and all were treated as threats to public safety. Francis seems to have been, if not obsessed with, at least unusually involved in Lowell's fire protection. Whether as an elected alderman (1849–50, 1862–64) or as a private citizen, his opinions on fire equipment carried a great deal of weight in the city. He kept detailed records of local fires in a large journal at the Locks & Canals office. Those that occurred in mill complexes drew his special attention.

Despite this spirit of mutual cooperation in firefighting, competition for water created tensions between the corporations and the city. Francis made it clear that the reservoir was a private structure, built at great cost (over $112,000) and requiring continual oversight and pumping. He was willing to let the city use its reservoir water in emergencies, but he was reluctant at first to provide unlimited access. Although the municipal government hooked up a few pipes to the private fire mains, Locks & Canals workers had to open their valves to fill those pipes whenever the city fought a fire. In 1852, Francis demanded, and got, a payment of $750 per year for this minimal service. Only in 1868 did the city expand its network of pipelines, gain control over the valves to the corporate water mains, and keep its pipes filled at all times. The price that Locks & Canals charged Lowell for this firefighting water rose sharply as a result, to $7,500 per year, or ten times the initial payment.

While Francis was helping to improve public safety (sometimes for a fee) and trying to avoid damage to corporate property, he was also finding ways to provide more power for the mill complexes that depended on the Locks & Canals system. In the winter of 1848, with work still under way on the three masonry-arched tunnels that comprised the Moody Street Feeder, he began to assess the improvements already completed. The Pawtucket Canal was under less strain now that a realigned dam funneled much of the river's water into the Northern Canal, whose Great River Wall was already perceived as one of the engineering wonders of the age. It was time to distribute the benefits from the purchase of the New Hampshire lakes and the expensive construction in Lowell.

A key issue was how to set new heights for the canal system. In 1846, Francis had warned corporations that Locks & Canals would probably raise the surface elevations in the canals. One of the principal purposes of the Northern Canal project was, after all, to bring water to the mills more efficiently, that is, with less friction head loss (drop in water level) from excessive current. It is clear, however, that not everyone on the system understood the full implications of higher levels of water at the mills.

The enlarged canal system could deliver water to the mills at a higher elevation, which would mean more power and should therefore be a good thing. But if Francis raised the controlled height of both levels, many of the installed breast wheels (some of which were new) might not work properly. Mills on the upper level rightly feared that the new setting for the lower level would flood their wheel pits and impede the rotation of their breast wheels. For example, at the Lowell and the Hamilton Manufacturing Companies, higher water in the lower Pawtucket Canal would force them to make costly changes in the position or form of their prime movers. W. C. Appleton, who had just completed a breast wheel project at the Lowell Manufacturing Company in 1846, complained that "the raising of the level of that [Lower Pawtucket] canal would be attended with the most ruinous results."[12]

After considering several schemes proposed by Francis, the directors voted in February 1848 to raise the controlled heights of water on both levels. The actual changes took some time, and a commission did not make final adjustments to the levels until 1853. The new elevations resulted in a gain of almost three feet of head for the Merrimack Manufacturing Company, which got the maximum benefit (going from thirty to thirty-three feet). The other mills on the upper level discharged into the lower level, and gained a foot of head (from thirteen to fourteen feet). Mills on the lower level, which discharged into either the Merrimack or the Concord rivers, gained two feet of head (from seventeen to nineteen feet). All these figures were approximate because water still lost some head as it moved through canals; a mill's position on the system mattered.

In 1851, a newspaper reported on the impact of raising the level of the lower Pawtucket Canal. This action would give more head to mills on that level but would "destroy, in a measure, the power of the breast wheels standing on the upper level." The newspaper added that "the breast wheels at the Carpet [Lowell Manufacturing Company], Lowell Machine Shop, and Hamilton Mills, are to be removed, and Turbine wheels substituted in their stead."[13] Nothing was said about the Appleton mills on the same level because they had been the first in Lowell to make the transition to turbines, which ran submerged.

Francis strongly encouraged mills whose breast wheels were adversely impacted by changes in canal levels to make the switch to turbines. Following his recommendation, the Hamilton Manufacturing Company ordered a huge Boyden turbine as a "substitute" for three of its many breast wheels in 1850. It was "the largest one of the kind (10 feet in diameter) every built in this city." The single turbine could produce 250 horsepower, over a hundred horsepower more than the three breast wheels combined, and it took up much less space.[14]

Damages paid by Locks & Canals for the injuries that manufacturers claimed as a result of the higher levels in the system included direct payments for that turbine as well as one at the Lowell Manufacturing Company and at the Lowell Machine Shop. Because the canal company was wholly owned by the corporations it served, these payments to three of the group were considered "joint expenditures" and part of the overall cost of "the improvement of the Water Power at Lowell."[15]

It was no coincidence that Francis had been promoting a transition from breast wheels to turbines since he started planning the

Northern Canal. Few have noticed this connection, but it appears prominently in his discussions of benefits from the northern lakes and the new canal. The seemingly cautious engineer was actually an energy revolutionary, pushing for change not only in the way energy was stored in New Hampshire and delivered by the Lowell Canal System but also in the way it was converted to mechanical power in the mills. His argument was persuasive: If textile manufacturers were going to have to adapt their waterwheels to take advantage of higher levels and additional flow in the canal system, then why not choose this moment to scrap obsolete breast wheels and switch to turbines that had higher efficiencies and behaved much better in backwater conditions?

The chief engineer promised that if the Boott mills, which discharged directly into the river, "were supplied with suitable water wheels, they can hereafter have water enough to give them as good speed in ordinary freshets as at any other time."[16] When Francis talked of "suitable water wheels" in this period he meant turbines. The efficiency of these wheels would not be significantly affected by backwater, and extra flow could compensate for a moderate reduction in net head.

In 1849, Francis designed and installed two inward-flow turbines for Boott Cotton Mills. That corporation was neither the first nor the last Lowell manufacturer to install turbines, but it probably had the most to gain. It may have had the worst position on the canal system when it joined the ranks of the major textile manufacturers in 1837. The site of the Boott mills, at the end of the lengthy Eastern Canal and right on the river, subjected them to both friction head losses and backwater. With only seventeen feet of head on the original lower level, as opposed to the thirty feet at the Merrimack mills next door, even modest levels of backwater were serious.

Adding the Northern Canal to the system did little for Boott Cotton Mills at first, but Francis was still developing ways to extend the positive impact of that new waterway. The Moody Street Feeder was designed to put substantial flow into the Merrimack Canal with minimal head loss. The feeder's outlets were curved, so water entered

Workers in the Merrimack Canal at the outlets of the Moody Street
Feeder Gatehouse in 1897. The task of removing sediments from drained
canals often took place on Sundays, when mills were closed. Courtesy of
the Center for Lowell History, University of Massachusetts Lowell.

the canal in the same direction as its current. As the directors of
Locks & Canals monitored progress on that underground structure
and its gatehouse in 1848, they were also considering the replacement of a tiny conduit that had been built in 1846 to bring extra
water to the Boott mills. The board authorized Francis "to construct a wooden penstock leading from the Merrimack Canal Wasteway to the Eastern Canal, for the purpose of supplying the mills [on
the] lower level with an additional supply of water, when they are
impeded with backwater."[17] Over time, the role of this "Boott Penstock" was expanded so that it could help maintain the higher level
in the Eastern Canal on a regular basis and help flush out ice that
built up there.

Because of the northern reservoirs, much more water could
be guaranteed in Lowell all year. The textile corporations wasted

little time putting that additional energy to work. They began to expand productive capacity even before Francis had worked out a fair way to distribute extra flow. Locks & Canals was then leasing only $91^{11}/_{30}$ millpowers, but demand was growing rapidly. Francis wanted the corporations to share the benefits of the improvements "in as nearly as practicable, the same proportions, as they own the waterpower."[18]

When the Northern Canal opened in 1847, there were thirty-four mills on the system. Within six years, there were seven more. Annual *Statistics of Lowell Manufactures* show that in the same period, the number of spindles rose from 253,456 to 342,722, and the number of looms from 7,756 to 10,606. Corporations began to use more waterpower in anticipation of a distribution of additional millpowers based on the percentage of ownership of existing leases. In general, those that already had the largest number of millpowers did see the greatest benefits, but the results were not exactly proportional. Extensive measurements that Francis completed in 1852 provided realistic indications of the amount of water that each corporation used for normal operations. Some companies were taking only a little more than their current leases allowed, but others, such as the Merrimack, Massachusetts, and Boott, exceeded the contracted amounts by a large percentage (see table 5.1). Most of the 51 percent increase in overall water consumption had come since the purchase of the New Hampshire lakes and the building of the Northern Canal.

The directors of Locks & Canals decided that they now had sufficient flow year round to meet the documented needs. In effect, the canal company did little more than charge the corporations for what they were actually taking. It then guaranteed those flow rates for the future. Initial redistribution of millpowers in 1853 closely matched the 1852 measurements. The division of additional powers favored corporations that were already large lessees of water. These decisions also reflected industrial demand and the need for balance on two levels. The purchase price of the new power was $3 per

TABLE 5.1 *Millpowers*

Company	Leases, to 1847	Tests, 1852	Leases, 1853	Flow (cfs), 1853
Merrimack Mfg. Co	$16\frac{15}{30}$	$24\frac{20}{30}$	$24\frac{20}{30}$	616.667
Hamilton Mfg. Co.	9	$15\frac{17}{30}$	16	968.000
Appleton Co.	$5\frac{15}{30}$	$7\frac{9}{30}$	$8\frac{16}{30}$	516.267
Lowell Mfg. Co.	5	$8\frac{12}{30}$	$8\frac{12}{30}$	508.200
Lowell Machine Shop	2	$3\frac{9}{30}$	$3\frac{9}{30}$	199.650
Middlesex Co.	4	$5\frac{23}{30}$	$5\frac{23}{30}$	262.383
Boott Cotton Mills	10	$17\frac{26}{30}$	$17\frac{26}{30}$	812.933
Massachusetts Cotton Mills	$14\frac{18}{30}$	$24\frac{16}{30}$	$24\frac{16}{30}$	1,116.267
Suffolk Mfg. Co.	$5\frac{10}{30}$	$6\frac{15}{30}$	$6\frac{15}{30}$	393.250
Tremont Mills	$5\frac{10}{30}$	$6\frac{15}{30}$	$6\frac{15}{30}$	393.250
Lawrence Mfg. Co.	$14\frac{5}{30}$	$17\frac{9}{30}$	$17\frac{9}{30}$	787.150
Total	$91\frac{13}{30}$	$137\frac{21}{30}$	$139\frac{11}{30}$	n.a.

Source: PL&C Directors, Apr. 1, Sept. 27, Nov. 1, and Dec. 17, 1853.

spindle (with one millpower equal to 3,584 spindles), a very reasonable amount. There was also a yearly fee of $300 per millpower.

One small discrepancy delayed the completion of the lease arrangements. Francis had found that the lower level used slightly more water than the upper. After the first redistribution, the Proprietors quickly negotiated the sale of an additional $1\frac{20}{30}$ millpowers on the upper level, raising the discharge enough to match the leased quantities on the lower level almost exactly. If we ignore the 616.67 cubic feet per second going to the Merrimack mills, which used the whole drop of thirty-three feet, we are left with 2,978.12 cfs leased on the upper level and 2,978.73 cfs leased on the lower level.

The final allocation divided $139\frac{11}{30}$ permanent millpowers among ten textile corporations and the machine shop. This represented an increase of $47\frac{28}{30}$ millpowers over the old leases. The millpowers assigned to each corporation were a minimum guarantee. For the time being, companies could also use surplus water most of the year without charge, as long as they created no problems with the operation of the canal system.

Before the new leases went into effect, most manufacturers were already improving their capability to use waterpower with the purchase of turbines. In January 1853, the Lowell Machine Shop and eight of the ten textile corporations had at least one turbine in their mill yards, although many breast wheels remained in use. When the Civil War began in 1860, only three textile corporations had any breast wheels left in their mills. There were then sixty turbines on the system, and most of them far surpassed the power of even the largest breast wheels.

Highly efficient turbines used less water for each horsepower they generated than did breast wheels operating under the same head. Francis was always seeking greater efficiency in water storage, water delivery, power generation, power transmission, and machine operation. He hated to see energy wasted. It must have been painful for him to watch water rushing over his dam in spring freshets, but that was nature showing him the limits of engineering control. Stopping leakage at the dam or through defective gates at the mills was a high priority in dry seasons. Occasionally his men had to spill water through wasteways to flush floating ice from the system, to keep proper levels in canals, or to supply extra demands by lower-level mills. He tried to keep that type of deliberate waste to a minimum. It is not surprising, then, that he turned his attention in 1857 to the one place on the system where waste had become standard practice: the end of the Merrimack Canal.

Since 1846, the Boott Penstock had connected the Merrimack and Eastern Canals; but the height of the Merrimack Canal was approximately thirty-three feet, while the Eastern was only nineteen feet. Supplementary flow of water to the Eastern gave up fourteen feet of head without producing any power.

Another issue that seems to have been on Francis's mind was the legal justification for the Pawtucket Dam. A series of "mill acts" in Massachusetts had determined that harnessing rivers for manufacturing was a public benefit. Locks & Canals had been a manufacturer from 1825 until 1845, when it sold the Lowell Machine Shop. Francis apparently believed that the canal company should get back into some kind of manufacturing to protect its rights. He asked for and received permission from the board of directors "to build and provide for the running of a mill" in 1857.[19]

The gristmill that Francis built at the end of the Merrimack Canal was in one way an economy measure: part of the waterpower that had been wasted at this site could now do some work. In other ways, it was an extravagance and a pet project of the chief engineer. This was by far the most ornate building erected by Locks & Canals and one of the handsomest structures in Lowell. E. C. Cabot, a prominent architect who had designed the Boston Atheneum, produced the elaborate color-washed drawings. They reveal the care given to a small manufacturing plant that Francis said was simply a mill for grinding corn. Its function may have been prosaic, but the building's Italianate/Romanesque architectural style and its prime mover were state of the art. Francis installed a large wooden turbine of original design to drive the set of French burr stones, corn cracker, and elevator in his mill. Locks & Canals was a water-powered manufacturer once again. Recreational walkers who strolled under the spreading elms along Dutton Street could pause to admire the handsome gristmill at the northern end of the Merrimack Canal. Almost as important to Francis as good construction was the landscaping that surrounded his company's buildings and waterways. As we have seen in previous chapters, he was following a corporate tradition that began with Kirk and William Boott.

Francis was truly an engineer with a green thumb. To him, designing with nature was part of the engineer's mandate. It was also a personal pleasure. His position as chief engineer gave him many opportunities to garden on a grand scale. He often took personal charge of the selection of trees and their placement on Locks & Canals property. His annual entries on the accomplishments of the past year almost always mention plantings. He specified types of trees and shrubs, showing considerable knowledge of arboriculture. In addition to purchasing large numbers of saplings from private suppliers, the canal company grew some trees in its own nursery near the Guard Locks.

Postcard showing promenaders on the Great River Wall of the Northern Canal. *Canal Walk Lowell, Mass* (Lowell, MA, n.d.). Author's collection.

The City of Lowell recognized the public value of corporate landscaping by giving Locks & Canals some relief on taxes for land that served municipal as well as corporate needs. When the city tried to tax the Shattuck Mall in 1858, Francis, in his position as agent, insisted on an abatement: "It is now & has been for about eighteen years, planted, fenced and kept in order by the Locks & Canals at their own expense as a public ornament of the City and on this account has not been taxed for many years until the present year."[20]

Francis encouraged people to enjoy safe parts of the canal system. A lengthy article in the *Lowell Courier* on January 1, 1848, predicted that the new Northern Canal would become the "finest place for promenading in the whole neighborhood." The chief engineer soon added a walkway along its entire length, from the Pawtucket Dam to the Suffolk mills. This path went through the island and across the top of the "Great River Wall," from which strollers had an unobstructed view to the western shore. The directors of Locks & Canals authorized iron fences along the route and paid for ornamental plantings, including shade trees.

Even before the walkway was fully landscaped, it was becoming very popular for all levels of Lowell society, at least on Sundays, when most working people had the day off. The *Lowell Advertiser* for July 14, 1848, said that "the New Canal has become a favorite resort for our citizens on the afternoon and evening of the Sabbath. It is a delightful promenade, and we are glad to see so general a turnout of all classes."

Sketches for "malls" that Francis intended to create on either side of the Northern Canal still exist. Intended for the inland section of the canal, these malls had double lines of trees. The note that accompanies his sketches is revealing of the man: "The above sections represent the form I want to have the banks when completed."[21] Here is an engineer thinking like a landscape architect, trying to design not just the shape of the earth behind each canal wall but also the form of the tree-lined malls that will beckon the citizens of his community.

Planting trees was clearly a pleasure for the chief engineer, but

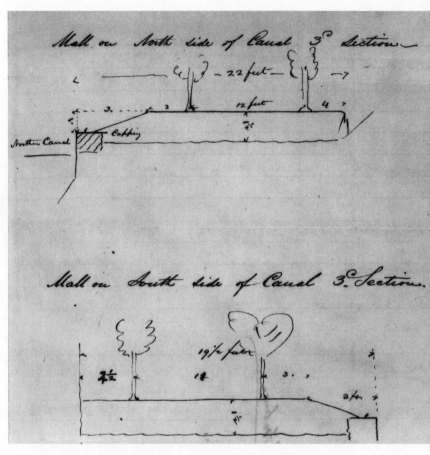

Sketches by James B. Francis showing the malls he planned in 1848 for both sides of the Northern Canal, along its inland section. Courtesy of Locks and Canals Collection, Baker Library Historical Collections, Harvard Business School.

the beneficial effects of landscaping efforts were not always immediate. Francis faced the same problem that perplexed Kirk and William Boott: the slow growth of the most admired shade trees. His solution along the Northern Canal was to plant an additional set of fast-growing trees (black Italian poplars) to provide immediate shade while more desirable elms were maturing. He cut the poplars down in 1859, when they began to crowd the elms.

Hikers could take a circuitous route out the Merrimack and Pawtucket Canals to the falls and back into the city on the Northern Canal. It should not surprise us that designers of lengthy canal systems would think of parks as systems and stress the extended or linked greenway more than the isolated green space. Walkways linked sections with different characters and offered promenaders on the Northern Canal a particularly striking sequence of visual experiences.

In the second half of the nineteenth century, Frederick Law Olmsted would become famous for planning such sequences at New York's Central Park, Brooklyn's Prospect Park, and Boston's Emerald Necklace. He would go on to become the nation's most influential advocate of park systems and greenways. Some historians go too far, however, when they suggest that he invented the idea of linear parks. Lowell already had two large city parks and a well-developed system of canal greenways a decade before Olmsted designed his first park in 1858, seventeen years before he proposed his first greenway, and more than a quarter of a century before he began planning an interconnected system of parks for Boston. The very active "greenway movement" of today may really have its origins in Lowell.

Trees were integral components of the canal system, but on one section of the greenway loop to and from the head of the falls, there was no room for these lush amenities: the narrow top of the Great River Wall was a bare walkway of stone, fenced in with iron railings for the safety of promenaders. Once people left the comfortable corridor of a tree-lined mall and started across that vertiginous wall, they could feel the full force of both the technological and the natural "sublime." Just beneath them on one side was the smooth but steadily moving water of the carefully engineered Northern Canal. On the other side, more than thirty feet below the artificial level of the canal, was the rocky bed of the Merrimack. On many working days, the only water in that natural channel stood in pools amid the crags of stone. Except during freshets, the canal system captured almost the entire flow of the river.

Streams like the Merrimack are valuable sources of power, but their natural flow is variable. The developers of Lowell were seeking relatively steady flow year round. In addition to developing factories and power canals, they worked to "improve" the drainage basin by building dams and storing water in reservoirs for later release in dry seasons. Their principal goal was to increase the "permanent" waterpower that was available in all but an exceptional drought. Another goal, which most historians have ignored, was to make productive use of additional, or "surplus," water. For much of the year, even a heavily engineered river carried a great deal more flow than the predictable minimum, or "permanent," quantity. Although most of that excess water went to "waste" (produced no power) by spilling over dams in freshets, it was possible to use a significant part of it for manufacturing.

In combination with the Essex Company of Lawrence, representatives of the Lowell corporations gained control of more than a hundred square miles of lake surface in New Hampshire between 1845 and 1859. They stored water in the lakes during the usually wet winter and spring and called for releases whenever the Merrimack needed additional flow to meet the downstream demands of industry.

With the new allocation of waterpower in 1853, Locks & Canals now had to supply a total of $139^{11}/_{30}$ millpowers to ten textile corporations and a great machine shop. Each of these permanent millpowers represented a specific flow of water (in cubic feet per second) that was expected to produce about 60 net horsepower on one of the drops in the two-level system. Technically, Lowell corporations leased water, not power. Their millpowers should be seen as contracts for the delivery of water with a certain amount of potential energy. The breast wheels or turbines, which converted that energy to mechanical power, did so with varying degrees of efficiency. By installing the latest improvements in turbine technology, mills often increased the net horsepower they generated from each millpower of water that Locks & Canals supplied.

The millpowers assigned to a corporation were only a guaranteed minimum. Lowell's millpond, which extended for eighteen miles above Pawtucket Dam, had to supply 3,595 cfs for fifteen hours of every workday to satisfy the leases for $139^{11}/_{30}$ permanent millpowers on two levels. Meeting that requirement was rarely a problem, even in the summer months. The annual flow of the Merrimack fluctuated but was usually close to the modern (1923–97) average of approximately 7,000 cfs. By storing all the water that came down the river when the mills were closed (nights and Sundays), Locks & Canals could theoretically satisfy all its contractual obligations with a river flow of only 1,926 cfs. In all but an extended drought, there was water to spare.

The vast millpond was a tremendous asset for Locks & Canals. It could hold almost twice as much water as the combined ponds created by the Amoskeag Company's dam at Manchester and the Essex Company's dam at Lawrence. All three of the Merrimack's water-powered industrial centers benefited from releases of lake water that moved down the river and into their ponds, but Lowell, with its greater storage capacity, had more flexibility. Francis could draw more water in working hours without immediate need of replenishment. Charles Storrow, the head of the Essex Company, was well aware of his relative disadvantage when he complained to Francis that "the pond between our dam and the foot of Hunt's Falls is so small in extent." He had to order extra water from the lakes when "your better means of holding in your pond render it unnecessary for you."[22]

Successful development and expansion of the canal system by the 1850s added luster to Lowell's already glowing national and international reputation. Attention was shifting from the city's efficient and well-dressed workforce, made up primarily of young women, to its tasteful architecture, bustling stores, shaded promenades, glistening waterways, and thriving industry. Lowell's water-powered mills, which had richly rewarded stockholders and provided employment for thousands, were also the largest source of municipal property taxes. Ambitious figures in other communities and planners of places not yet created were envious of Lowell's prosperity and eager to emulate strategies that had worked so well. Water rights and property at substantial waterfalls were now highly valued assets.

Waterfalls had always been favored places to create settlements. Both Indians and colonists considered them ideal spots for taking fish and for engaging in trade. Some falls were the end points of navigation from the sea or serious obstacles for river travel from the interior. Geographers have suggested the "break-in-transportation hypothesis" as one explanation for urban development. Enterprising companies used locks and canals to move vessels and rafts around drops on rivers. Others linked railroads to these busy points. One of the most important assets of rivers like the Merrimack was abundant waterpower, which could turn the wheels of industry. Lowell proved that large-scale waterpower development was not only technically feasible but also lucrative.

Boston investors took the lead in applying lessons from the Lowell experience at other northern New England waterpower sites, including Chicopee, Lawrence, and Holyoke in Massachusetts; Nashua, Manchester, Dover, and Great Falls in New Hampshire; and Saco and Lewiston in Maine. These communities, although different in various ways, were part of the growing industrial operations and urban development efforts of the Boston Associates, who would either buy out struggling operations or acquire undeveloped sites. They all depended heavily on waterpower, and all but Holyoke, which turned increasingly to paper production and used a great deal of process water, focused on textiles.

Many of the Boston Associates, hoping to repeat the success of Lowell in other places, freely borrowed both engineering and planning concepts. In 1836, George White said that Cabotville (in Chicopee, Massachusetts) was a "pleasant village" that was "growing up with astonishing rapidity, and bids fair to become at no very distant day, a second Lowell."[23]

Planners who had lived or worked in Lowell played critical roles in building new cities and waterpower systems at waterfalls. Ezekiel Straw laid out most of Manchester for the Amoskeag Manufacturing Company in the late 1830s, improving an existing two-level canal system parallel to the Merrimack River. In the following decade, Charles Storrow was responsible for the initial design of Lawrence, where a single canal avoided the difficulties of balancing flow on several levels. Relatively straight stretches of river at both sites meant that canals and streets could be placed parallel to the channel. The water's edge was, of course, reserved for mills, with rows of workers' housing not far away. Orderly grids and space for public parks were priorities in the rational plans for these cities. In their urban forms, they were not close copies of the Lowell model, but the waterpower potential at all three sites was similar.

Good urban planning and available waterpower did not bring immediate economic success to all the Boston Associates' ventures. Lawrence was slow to find lessees for all its permanent waterpower, and exploitation of immigrant workers tarnished its image. At Holyoke, where the first dam collapsed right after construction in 1848, a rebuilt Connecticut River dam and a three-level power system offered more waterpower and mill sites than there were buyers. The 1880 census said that "nearly all the permanent available power is utilized" at Lawrence, but "considerable power [is] still available" at Holyoke.[24]

Lowell never had any problem leasing all its permanent millpowers, and its industrial growth seemed to have no limits. The city was a model not just for the Boston Associates, and not only in New England, but for promoters of industrial communities large and small all over America as well. In 1828, one of the Quaker owners of the Brandywine Mill Seat Company in Delaware told E. I. DuPont about an effort by Philadelphia investors to buy them out: "W. Young says they wish to establish a Lowel concern, which I expect thee has read of."[25]

When urban boosters invoked the name of Lowell like a talisman, they hoped its aura would steel their resolve and persuade others to invest in their dream. There were justifiable hopes for Minneapolis, where the mighty Mississippi spilled over a spectacular cataract. Some residents had actually wanted to name the place *Lowell.* The famous city on the Merrimack also helped to inspire significant waterpower development at the great falls of the Mohawk in Cohoes, New York, where a six-level canal system may have set a record for hydraulic complexity. Even when a city government took on its own waterpower projects, as at Augusta and Columbus, Georgia, the Lowell ideal was inspirational.

By 1856, Francis was ready to tinker with the methods of water delivery and billing that had helped to make Lowell a model for other waterpower providers. Certain that he could supply more than the permanent millpowers for most of the year and confident that he could determine how much surplus flow was going to each mill complex, he proposed that rent should be paid for that extra water. The directors of Locks & Canals took no immediate action on his suggestion, but Francis continued to collect data in support of his arguments. According to the indentures (leases) of 1853, manufacturers on the system had the right to use surplus water under terms and conditions that were to be set by Locks & Canals. For the next six years, Francis allowed mill superintendents to take additional flow (above the amounts in their leases) without charge, as long as they created no serious problems with the operation of the canal system. The chief engineer was, however, already thinking of ways for Locks & Canals to control the distribution of this surplus and make a profit in the process.

Using methods proven accurate by testing, Francis started to keep close track of water consumption in 1856. He quickly established the regular use of flumes and tubes to measure water use at selected mill complexes. He was also monitoring turbine speed-gate settings in some mills and relying on several wasteway weirs or apertures (openings through which water flowed) to give him additional information. In 1858, he suggested that the directors of Locks & Canals re-examine the "entire system" of measurement. "These operations as you are aware are attended with great expense." It was important that the directors decide "whether to continue the present precise, but expensive system, or to adopt some other method, less exact and costly."[26] Francis left the decision to them, warning that some way of monitoring would be essential once summer began. He predicted water shortages that summer if individual corporations were allowed to take all the water they wanted without any checks.

The directors, who represented the corporations that depended on waterpower, decided to stick with Francis's "precise, but expensive system." It was not necessary to imply that some mill complexes would take advantage of the situation if they were not carefully monitored or that even the best-intentioned superintendents might inadvertently take more water than they should. Measurement at that time was simply a way to keep the canal system in balance, to promote equitable distribution, and to avoid shortages in dry seasons.

The use of surplus water had been slowly increasing in the 1850s. By 1859, the mills were taking approximately 40 millpowers in addition to the permanent leases of $139^{11}/_{30}$ millpowers. The surplus was both a potential source of income and a threat to the effective management of the canal system. Up till then, the only tool that the chief engineer possessed for regulating surplus was his power to prohibit its use entirely.

Francis warned the president of Locks & Canals that he needed a better way to restrain the consumption of surplus water. Demand for power was rising, and the capacity of the canals was limited. He wanted to charge for surplus at "a very high rate, as high as it will bear, irrespective of the cost of the permanent powers." Francis determined that the average cost (considering both the initial purchase price and the annual rental) of a permanent millpower of water in Lowell was about $3.75 per working day. The cost of an equal amount of steam power in the city was much higher by his estimates: $9.71. Francis then asked, "Would half the price of steam power be too much to charge for the surplus water power?" He concluded that $5 per millpower per day was an appropriate fee.

The chief engineer also requested power to stop "the use of part of the surplus as well as the whole."[27] Francis wanted more flexibility in setting limitations.

On April 12, 1859, Locks & Canals established rules and regulations for selling surplus water. The canal company was careful not to concede any "permanent right or privilege whatever," and it retained the option to stop all use of surplus.[28] Manufacturers had to pay only for the additional water they actually used, with billing beginning at $3.50 per millpower per day. To encourage restraint and prevent too much dependence on surplus water, the fee for any excess over 30 percent of the leased amounts jumped sharply to $7. In June of the following year, the directors amended the regulations so that Francis would have the authority to set specific limits, depending on the availability of water. As representatives of all the corporations, they wanted to treat everyone fairly and to avoid any sense of favoritism. A limitation would be the same percentage (25 percent, ten percent, etc.) of each corporation's permanent millpowers. Francis had the authority to change that maximum percentage of surplus water at any time. Whenever there was not enough water for every corporation to have some extra flow, he would prohibit surplus for all.

In order to charge for surplus water by the day, Locks & Canals had to measure regularly. This was more than a case of simply catching offenders who took more than their permanent millpowers allowed. In theory, Francis now needed to know how much every corporation was using, every working day. He encouraged mill superintendents to tell him if there was any change in their normal operations, such as a breakdown of machinery or a special draft of water for some process. He also let them know that he would be sending teams to their complexes to check flow and monitor gate settings. Nothing in the leases limited his rights to measure or to inspect.

Wooden flumes that were one hundred to one hundred eighty feet long and usually as wide as a canal had become permanent fixtures in many of the waterways by 1856. Although expensive to build and maintain, these structures did not obstruct water move-ment or affect head. One of their great advantages was that they measured all the flow going to a mill. Francis used basic displace-ment theory to calculate how much lead should be used to sink each sheet metal tube almost to the bottom of a particular flume. The fabricator then soldered a specified cylinder of lead to the lower end. At the flume, his men poured in lead shot as needed for fine adjustment. A cork capped the loaded tube, which extended about four inches out of the water while in use. A scale at the top indicated its submerged length. That length and the depth of the flume were important factors in his correction formula for determining flow.

The usual practice was for teams to take four sets of measure-ments per week in each flume. In droughts they might do even more. Three of the chief engineer's sons, James, Joseph, and Charles, worked on flow measurement, which was an attractive assignment for young men with engineering aspirations.

It was essential for Locks & Canals that textile executives have faith in the accuracy of Francis's water measurements. He earned their trust because he did rigorous tests to prove the validity of his gauging procedures. They knew that he published accounts of his experiments and invited comments from prominent people in his profession. Perhaps most important was his reputation for being honest and responsible. He went out of his way to correct even the slightest mistake. On one occasion, Francis saw what he thought was a slight curve in the upstream face of a transit beam across one of his permanent flumes. After careful surveying, his men deter-mined that it was indeed deflected by exactly 0.142 feet in the mid-dle. He immediately had recent flow figures recalculated to allow for the curve and then fixed the beam.

Locks & Canals expected its customers to buy surplus water at the minimum rate rather than turn to the alternative of steam power. Waterpower was almost always less expensive than steam power at sites where nature cooperated to provide a drop on a river with reli-able year-round flow and where it was not too costly to build a dam and power canal system. Normal operating expenses were much lower than the continuous costs of buying fuel for steam power. Waterwheels or turbines were not cheap, but neither were efficient steam engines, such as the ones that George Corliss made popular in the second half of the nineteenth century. Unlike turbines, steam engines required almost constant oversight by skilled operators and a great deal of maintenance. Once manufacturers had made the capital investment to harness waterpower at a site, their costs to maintain that source of energy were relatively low.

It is significant that we find almost no mills scrapping their breast wheels or turbines to run steam engines in their place. John R. Freeman observed that there were factories paying companies like Locks & Canals for waterpower while they had "emergency steam engines standing idle."[29] Certainly there were places where waterpower was not a possibility or was too expensive to develop; waterwheels had more locational constraints than steam engines. There were some mills that had free fuel for steam boilers (woodworking establishments could produce piles of burnable scrap and sawdust), and a special need for process heat from exhausted or condensed steam might make an engine unusually economical to operate. In cases like these, steam power was a logical choice for an initial installation or an expansion, but it was seldom a good *replacement* for waterpower.

As early as 1839, Patrick Tracy Jackson was suggesting that Locks & Canals should provide "sites for mills to be supplied with water when the river is full, say 8–10 months in the year." He never mentioned steam power, but, even in 1839, that would have been the obvious way to run these mills when no surplus water was available. The new mills were not to have any permanent, guaranteed millpowers of their own. Although nothing came of this idea, Francis revived a similar scheme for a "power mill" in the early 1850s. He was explicit about the mix of power sources in this building, which would be leased. The mill was "to be driven by water whenever we have it to spare for that purpose, the remainder of the year say six months to be driven by steam."[30] Again, Locks & Canals did not implement the plan, but corporations *already* on the system did take advantage of both surplus water and steam power as they expanded their plants.

The treasurers of the Lowell textile corporations wanted to keep Locks & Canals waterpower restricted to their own exclusive group, all of whom focused on large-scale manufacturing. They would not rent space with power to smaller firms, a practice that helped build a diversified and dynamic manufacturing base in other Massachusetts cities such as Fall River and Worcester. The focus in Lowell was on expansion of productive capacity by the great mill complexes, using all the waterpower available and any additional steam power that was needed.

At the best waterpower sites, and Lowell was one, there was little reason to favor steam over the natural energy from a fall, but it was often advantageous to use steam as a *supplementary* power source. Most water-powered mills suffered from occasional power shortages in dry seasons or in freshets, and some simply wanted to expand production but had already reached the limits of their available waterpower. These were the companies that could benefit most from steam power, either as an emergency backup or as part of a hybrid (steam/water) power system.

Supplemental or auxiliary steam power had long been used in combination with waterpower for manufacturing. In Britain, there are examples in the 1740s and 1750s of Newcomen steam engines pumping water back up to vertical waterwheels, which then produced rotary power. This practice could artificially augment or replace stream flow in droughts, provide a continuous source of additional power for expanded production, or function in a closed system not connected to a stream. Later in the century, engines by Watt, Smeaton, and others joined Newcomen engines in pumping for waterwheels. This particular type of steam/water combination became obsolete after Watt's improvements in the rotative engine between 1780 and 1800.

In mills during the nineteenth century, steam engines could assist, fill in for, or completely replace waterwheels and turbines. The use of auxiliary steam engines became a common practice in both Britain, where available waterpower was limited, and in the United States, which had numerous falls with relatively steady and abundant flow. Such hybrid systems were also used in France, but to a much lesser degree. In the American textile industry there were many combined steam/water mills, including the Wilkinson Mill in Pawtucket, Rhode Island, which had both a steam engine and a breast wheel when it opened in 1811. Hybrid systems could keep a factory running despite occasional shortages of waterpower (during droughts, heavy ice conditions, or floods). Their boilers and engines provided not only mechanical power but also exhaust steam, which was valuable for heating (and sometimes humidifying) workspaces and for textile finishing processes such as printing and dyeing.

Corporations on the Lowell Canal System felt more secure when they had reliable steam power in reserve. They also saw steam as a way to expand their operations when demand for textile products increased. Even after system improvements and lake storage made possible additional leases and sale of surplus power, most manufacturers quickly found use for all the water available in dry seasons. If they either wanted to keep up a high level of production all year or intended to enlarge their mill complexes, they had to build their own steam-power plants. Beginning in the late 1840s, more and more corporations began to depend on some steam power, if not all the time, then for at least part of every year.

Efforts to divert as much water as possible into the canal system were growing more costly for wild creatures that had formerly navigated the natural falls without undue difficulty. The number of anadromous fish caught in the Merrimack River at Lowell had declined sharply in the decades after Kirk Boott built his temporary dam in 1825. That dam and additional dams on the main stream and tributaries were particularly damaging to the Atlantic salmon, which had once been "very abundant in the Merrimac river." Dr. David Storer, an expert hired by the state, reported in 1839 that "the building of dams and manufacturing establishments, by preventing the fishes from going up the rivers to deposit their spawn, has almost entirely annihilated this species in our state."[31]

Writing about his own 1839 trip on the Concord and Merrimack, Thoreau provided a sad commentary on the plight of another anadromous fish, the shad: "Still patiently, almost pathetically, with instinct not to be discouraged, not to be *reasoned* with, revisiting their old haunts, as if their stern fates would relent, and still met by the Corporation with its dam. Poor shad! where is thy redress? When Nature gave thee instinct, gave she thee the heart to bear thy fate?"[32]

Massachusetts legislative acts that were intended to maintain fisheries and remove obstructions to the migration of salmon, shad, and alewives did little to improve the situation on the Merrimack River before the 1860s. When the Essex Company acquired a charter in 1845 and, with it, permission to build a tall dam at Lawrence, the state required the company to maintain a fishway at the structure. Not only did the fishway not work, but floods swept it away several times. After much litigation, and payment of damages to fishermen, the Essex Company finally defeated an 1856 act that would have forced it to build an effective fishway. In 1859, Chief Justice Shaw ruled (*Commonwealth v. Essex Company*) that the state had allowed the company to compensate fishermen for their losses and that those payments relieved it from any responsibility for providing a fishway. His decision, which relied on contractual arguments to declare the 1856 act unconstitutional, had serious implications for Massachusetts and New Hampshire fisheries. Not even the strongest fish could jump high enough to clear the thirty-two-foot face of the Essex Company's dam.

Locks & Canals had maintained a fishway on its much lower dam since 1830. However, the malfunctioning fishway on the dam in Lawrence set a bad precedent. Its ineffectiveness convinced the corporations in Lowell that there was no point in letting valuable water escape for the benefit of fish that were already blocked downstream. On December 26, 1854, Francis noted in his record book that "the fishway in the main dam was permanently stopped up."[33] Despite protests by fishermen, it would remain that way until after the Civil War.

A watertight dam, a new canal, storage reservoirs, and more millpowers raised expectations for higher profits among the textile corporations. Although everyone got more waterpower, some mill complexes were still in better locations than others. Only a few years after the 1853 leases had gone into effect, inequities in the system again became a heated issue. To alleviate the problems, Francis proposed two additional feeders from the Western to the Merrimack Canal and a new channel paralleling part of the Pawtucket Canal. His directors decided that these projects were too costly in a time of worsening economic conditions.

Lowell weathered the national financial crisis known as the Panic of 1857 but was not prepared for the Civil War, which sent prices of raw cotton soaring in 1860. Overly cautious textile corporations in the city made little effort to find other things that they could make while cotton was in short supply. Unlike Manchester, where the Amoskeag Company increased its sales of fire engines and took large federal contracts for muskets, Lowell seemed paralyzed. Some of the great mill complexes actually shut down all their manufacturing. However, confidence in the future of cotton products remained high. Corporations took advantage of the slowdown to add new mills, enlarge or link existing buildings, and continue to improve their prime movers. At Locks & Canals, Francis soon found that he could undertake modifications of the canals that would normally interfere with large-scale textile production, such as the much-needed enlarging of the Merrimack Canal from the point where supplemental flow from the Moody Street Feeder created high currents and head loss.

Francis was never finished with repairs or incremental improvements to the canals. Like most complex technological systems, the network of waterways and control structures was a work in progress. Even the goals that the chief engineer established were moving targets, strongly influenced by political events, economic trends, technological advances, and manufacturing priorities.

As the long and bloody Civil War came to an end, Lowell's great cotton textile producers were poised for a surge in fabric sales and profits. Mill enlargements and new construction during years of little or no manufacturing had expanded productive capacity. Highly efficient turbines, which worked well even in backwater, had replaced almost all of the old breast wheels. Devastating floods were less of a threat since the installation of Francis's Great Gate. Fire protection in both the mill yards and the surrounding community was exemplary. The expanded canal system of the Proprietors of Locks and Canals was not perfect, but it could supply the leased millpowers, as well as a great deal of carefully monitored surplus water. And steam engines were already becoming an important part of the power mix in many corporations, offering reliable energy that was unaffected by freshets or droughts.

CHAPTER SIX

Controlling the System, 1865–1885

The Civil War had ended and Lowell was thriving again. The sky was still dark when drops began spilling over flashboards on the crest of the dam. A chilly breeze ruffled the fall foliage of riverside trees. Water had been rising slowly since the workday ended in Lowell's textile mills at 6:30 the previous evening. The dam created a millpond that extended eighteen miles up the Merrimack River, beyond Nashua, New Hampshire.

The first sound of mill bells came at 4:30 AM, announcing that the workday was near. At the same time, employees of the Proprietors of Locks and Canals raised their sluice gates to tap the impounded water that fed their power system. When the third of the morning bells sounded at 6:20, operatives were already crossing canal bridges on their way to work. If they looked down at the water, they would have seen brightly colored leaves floating toward the mills.

At 6:27, three minutes before production was scheduled to begin, laborers at the ten manufacturing corporations manually opened speed gates that set most of their fifty-seven water turbines in motion. Young women in long skirts hurried up flights of stairs and across oil-stained floors to their machines, while overseers checked their watches and paced impatiently back and forth.

The rippled surface of the pond had stopped rising and was slowly starting to fall, for its water was now rushing into a five-mile network of canals. The great industrial city was taking more water per second than was flowing down the Merrimack from the northern tributaries and lakes that fed it. In the mills, turbines had reached full speed. Leather belts slapped against pulleys on the spinning line shafts. At exactly 6:30, thousands of mill operatives threw the levers that linked their machines to the power of the river. Another day of making textiles was under way.

Peace in 1865 brought resumed shipments of southern cotton for the reawakening mills of Lowell and made control of both the canals and the New Hampshire reservoirs more complicated. As a group, the textile corporations now had more spindles than before the war and were more efficient in their use of each cubic foot of water. Steam engines were rapidly becoming necessities, for either supplemental or permanent power. In order to keep all their machinery in operation, managers were willing to pay for surplus water when it was available or for extra coal when it was not. James Francis was advising the corporations on how they should make use of both water and steam. He would have to keep the dam, gatehouses, and waterways in good shape and the system in balance. Competition from other manufacturing centers was increasing, but waterpower at reasonable cost still gave Lowell an advantage over places that depended entirely on steam power.

Most of the textile corporations had an effective strategy to power their enlarged operations: they depended heavily on their leases of "permanent" waterpower, bought as much surplus water as they could afford, and used only as much steam power as they needed to maintain full production. This strategy helps to explain how Lowell could continue to enlarge its manufacturing base in the face of growing competition from a coastal city like New Bedford, which had no waterpower but received its steam coal by relatively inexpensive maritime transport.

Workers in the mills bore the brunt of other cost-cutting measures. Corporations held down wages, sped up the pace of work, and eliminated some of the paternalistic practices that had given Lowell its utopian aura. Even before the shutdowns and layoffs of the wartime years, Irish women had taken many of the jobs formerly held by "Yankee Mill Girls." Although New England women returned in force at the close of the war (some having lost husbands, brothers, or fathers in the conflict), the number of immigrant employees increased over time. Some of the company directors no longer saw a need to actively recruit native-born operatives for their workforce or to maintain a model industrial community. Local managers still cared about the appearance of company property, but profit and higher levels of production had become the highest priorities.

Continuing increases in the number of spindles and looms created more demand for waterpower in Lowell. The directors of Locks & Canals were concerned about unbridled use, which could overwhelm the capacity of their system. In an apparent attempt to restrain the consumption of surplus water in 1866, they doubled the modest fees that had been set in 1859. Flow-measurement teams made precise determinations of the amounts of water that heavy users drew from the canals.

For the rest of the decade, the higher fee structure for surplus water remained in place. Meanwhile, some manufacturers on the canal system were investing heavily in increasingly efficient steam engines. The directors of Locks & Canals recognized that steam power was an important supplement for waterpower, but they did not want to see manufacturers choosing to run steam engines instead of turbines when extra water was available. The costs for steam power were falling because of lower coal prices and higher efficiencies in steam engines and boilers. In 1870, the directors encouraged reasonable use of surplus water by cutting the minimum charge to $5 per millpower per day and the charge at the next level (31–40 percent of the leased power) to $10.

This step may have been necessary to keep the price of surplus water attractive. On the other hand, preventing excessive water consumption was still essential. The delivery system did not work well when too much surplus water was moving through the canals. As we have seen in previous chapters, the friction caused by high currents wasted energy and reduced head. At the same time that they cut fees for reasonable use, the directors of Locks & Canals decided to raise the fees for taking more than 40 percent of the leased quantity and to heavily penalize any corporation that exceeded a 50 percent ceiling. There would be no mercy for anyone who exceeded a special limit set by the chief engineer when water was scarce.

Labor law had a direct effect on surplus policy. In 1874, the state of Massachusetts passed a law that effectively limited the working day in textile mills to ten hours. The textile corporations opposed the ten-hour law for its impact on production, but this legislation did make it easier for Locks & Canals to meet its minimum flow requirements and even to provide some surplus almost all year. Lowell mills, which had been running eleven hours per day, would now be using water for less time each day, and more flow would therefore be available. The directors of Locks & Canals carefully examined the projected impact of the law on water consumption and then raised the percentage limits for each price level "during the operation of the ten hour law."[1] In 1881, Francis remarked that he had been forced to prohibit all use of surplus water only forty and one-half days since 1874.

Locks & Canals always provided an important exception in times of natural backwater, when mills along the river had to take more than their leased flow to get the minimum power they needed to operate. Since freshets impeded the rotation of breast wheels and reduced the effective head for turbines, mills got less mechanical horsepower from each millpower of water. Only by using more water could they make up the gap. At first, the canal company charged nothing for surplus water during backwater conditions (established by a height gauge in the river behind the Merrimack mills). In 1875, the company established a nominal fee of one dollar per millpower per day.

Surplus water fees were a sharp departure from the normal practice of purchasing rights to and then paying an annual rent for permanent millpowers. Water had clearly become a commodity when payment was made only for what was used on a given day. With permanent millpowers, customers paid to take a specified flow of water during every workday. If they did not need the contracted water, they would still have to pay the rental charges or risk losing their allotment. Surplus water was different: it was an option, available on demand, up to the limits set by Locks & Canals.

Although the canal company, which was wholly owned by the corporations it served, was not in business to make a profit, extra income from surplus water sales was welcome. The annual rent for each of the $139^{11}/_{30}$ permanent mill-powers (after paying for the initial leases) was only $300 per year. That amounted to $41,810 per year. Francis said that almost all the money was "expended in maintaining and improving the water-power."[2] Surplus fees apparently made possible many of the special projects that enhanced Francis's sterling reputation in the engineering profession. His meticulous work on hydraulic and mechanical experiments continued throughout his career. Locks & Canals heavily subsidized the publication of three revised additions of his *Lowell Hydraulic Experiments* between 1868 and 1883.

Taking regular measurements of flow was an expensive, labor-intensive process, but Francis could justify it by pointing to the money earned by surplus water fees. An entry for fiscal year 1881 listed $9,551.70 as the total cost of measuring water. The recorder subtracted that amount from the $29,152.21 paid by the corporations for extra power. The provision of surplus water netted $19,600.51 for Locks & Canals. That was a substantial figure, almost half as much as the annual income from renting permanent mill-powers.

There were tradeoffs for this additional money. Taking too much surplus water could create problems for both the canal company and its customers. Heavy demands by just a few mills could upset the dynamic equilibrium in the canal system. All but one complex on the upper level discharged into the lower level. In order to prevent waste, Francis tried to keep the demands on both levels in balance. Another continuing problem was friction head losses caused by excessive current in the canals. Reducing the velocity of water in the Pawtucket Canal had been one of the principal reasons for adding the Northern Canal in 1847. Manufacturers who drew surplus water caused higher currents, and thus more friction, in the canals.

Mill superintendents were glad to have the additional flow, but they were dismayed by the diminished head. There were frequent situations in which water arrived two to three feet lower than the established level. The higher currents also created difficulties for navigation in the canals, which still functioned as transportation corridors as well as conduits for power. Although much less cargo moved through the canals than in earlier periods, Locks & Canals was still bound by law to maintain the transportation capability.

Finally, there was the danger of releasing too much water from the New Hampshire reservoirs too soon, leaving an inadequate reserve for a possible drought in the fall. Francis was very careful when he drew down the lakes that fed the Merrimack River. Writing in 1875, Samuel Lawrence said that "it is a most fortunate circumstance that the care of these lakes has been entrusted to a man who has no equal in that branch of science, to which his life has been devoted."[3] Francis kept detailed records of lake levels and daily rainfall, but it was even more difficult to predict the weather in New England in the nineteenth century than it is today. Whenever there was any doubt about maintaining sufficient storage, he put strict limits on surplus water. Everyone agreed that it was essential to assure enough flow for the permanent leases.

Clemens Herschel, who operated the canal system at Holyoke, adopted similar rules for surplus water. He was a close associate of Francis and shared his concerns about companies that took too much. At a birthday dinner, a friend reported that Herschel "restricted the use of water at times of low flow and penalized by ten times the regular rate any who exceeded their allotment. An irate customer, after fruitless expostulation, quoted some lines about being 'crowned with a little authority' and making angels weep. Mr. Herschel rejoined that if the angels wept any surplus tears they would have to pay for them."[4]

Companies usually installed much more turbine capacity than was necessary to handle the flows guaranteed in their leases with Locks & Canals. Turbines could operate under a wider range of both head and flow than could breast wheels. The choice of a more powerful turbine might not increase the purchase price a great deal or require a significantly larger wheel pit. Any of the mill superintendents ordering new turbines would seek enough capacity to run all their machinery even when there were a few feet of backwater, and most of them wanted to take advantage of surplus power whenever it was available at a reasonable price.

It is difficult to assess the exact power that mills on the system really produced at any point in time, but we do have an engineering report from 1883 that compared the capacity of installed wheels at each corporation with the amounts of water in their permanent leases. Only the Lowell Manufacturing Company, a maker of carpets, had turbines that could not handle the leased flow. Every other manufacturer had excess capacity. The mean ratio of capacity to leased flow was 1.80. Four of the largest corporations (Merrimack, Boott, Lawrence, and Massachusetts) could accept more than twice the contracted amount of water.

The heavy use of surplus water during backwater conditions is apparent in statistics that Locks & Canals compiled for the period from 1875 through 1884. Mills suffered from backwater an average of 111 out of the 309 days working days in every year. The corporations as a group used an average of 60.23 surplus millpowers in backwater, compared with 24.62 surplus millpowers when there was no backwater (198 days). The average number of surplus millpowers in use on the canal system, regardless of river conditions, was 37.37, or 26.8 percent of the total leases.

Changes in the river, the canal system, and wheel efficiency had gradually increased the net horsepower equivalent of a millpower from less than 55 to more than 70 by the mid 1870s. Using the latter figure as a conservative estimate of one millpower, we can calculate that from 1875 to 1884, surplus water (when there was no backwater) provided an average of 1,723 horsepower in addition to the 9,756 horsepower from the corporations' $139^{11}/_{30}$ permanent millpowers. The total waterpower from the system was therefore almost 11,500 horsepower in normal conditions. The mills took much

TABLE 6.1   *Comparison of Leased and Installed Waterpower in 1883*

Mills	Leased millpowers	Equivalent leased horsepower*	Ratio installed/ leased	Equivalent installed horsepower*
Merrimack	24⅔	1,727	2.14	3,695
Hamilton	16	1,120	1.76	1,971
Appleton	8⁸⁄₁₅	597	1.58	944
Lowell	8⅖	588	0.90	529
Middlesex	5²³⁄₃₀	404	1.34	541
Suffolk & Tremont**	13	910	1.47	1,338
Lawrence	17³⁄₁₀	1,211	2.18	2,640
Boott	17¹³⁄₁₅	1,251	2.04	2,551
Massachusetts	24⁸⁄₁₅	1,717	2.01	3,452
Lowell Machine Shop	3³⁄₁₀	231	2.59	598
Total	139¹¹⁄₃₀	9,756		18,259

Sources: PL&C Directors, Dec. 17, 1853, and appendix, Shedd & Sawyer Reports, 1883, A17, #83, PL&C-Baker.

*For this table, each millpower in 1883 is considered to be equivalent to 70 net horsepower.

**The Suffolk and Tremont Mills were merged by this date.

more water in backwater conditions, but we can't tell how much power they actually realized.

The 1880 U.S. Census said that "during 'backwater' large amounts [of surplus] are used, some companies using up to 75% of what they own [permanent millpowers]."[5] Although we know how many millpowers (or how much water) the mills as a group used in backwater conditions (from the 1875–84 data), we do not know what mechanical power they were getting from that water. The *level* of the backwater varied constantly, and thus the average head (essential for calculating waterpower) is impossible to determine. It is safe to say that backwater was primarily a problem of the lower-level mills along the river and that it seldom halted production after turbines had replaced breast wheels. Most mills that used a lot of free or very cheap surplus water during moderate backwater conditions were probably getting more than enough power from it to compensate for any loss of head.

The program of surplus water sales had many environmental implications. Although it is tempting to credit it with reducing the burning of coal in Lowell (by providing waterpower, which was a cheaper alternative to steam power), that was not the case. Technology frequently causes unexpected outcomes. Some corporate decisions to install steam boilers and engines for textile production were influenced by the availability of surplus water most of the year. Much of the steam power on the canal system after the Civil War was supplemental, not primary. The paradox is that without surplus water, Lowell's textile corporations might not have invested as heavily as they did in steam power.

Surplus water was a significant stimulus for both technological change and industrial expansion in Lowell. For this unreliable source of extra waterpower to be most useful in textile manufacturing, however, it had to be combined with the capability to generate steam power. The 1848 *Hand-Book for the Visitor to Lowell* had spoken of a large amount of "variable" waterpower, which, "with the addition of steam, might become available." For those mills that did make use of the surplus water, only the installation of steam boilers and engines could assure continued full production whenever the river had sharp reductions in flow. The promise of substantial amounts of surplus water for most of the year made it worthwhile for corporations on the canal system to add steam engines and textile machinery. Extra water could cut the consumption of fuel and provide major savings for industrialists in interior cities like Lowell, where coal was delivered by rail.

In Lowell, the purchase of steam engines and the burning of coal (producing air pollution in the city and mining wastes in Pennsylvania) increased sharply after the Civil War. Although engines had been in use for power (and their exhaust steam for processing) at the Merrimack Print Works, the Lowell Manufacturing Company, and Middlesex Manufacturing Company since the late 1840s, the annual *Statistics of Lowell Manufactures* provided no data on steam power until 1866, when they listed a total of twenty-nine engines,

TABLE 6.2   *Steam Power and Spindles on the Lowell Canal System*

Year	Spindles	Steam engines	Installed horsepower
1865	437,420	29	2,885
1870	499,806	30	3,690
1875	692,888	67	7,733
1880	703,670	88	11,950
1885	834,100	161	20,600

Source: Statistics of Lowell Manufactures (Lowell, MA, 1866, 1871, 1876, 1881, 1886). Spindles at the textile mills on the system are one measure of productive capacity. These figures do not include steam engines at the Lowell Machine Shop, a complex that did not make textiles.

with 2,885 horsepower, on the canal system. By 1871, there were thirty engines with 3690 horsepower, and ten years later, eighty-eight engines with 11,950 horsepower. It is important to remember, however, that not all of this power was in use on an average workday. Surplus water often kept steam engines in reserve.

Surplus water actually gained in value once corporations could buy efficient steam engines, like those made by George Corliss. That may seem illogical, but not if you understand the demands of cotton textile production. Each of the mill complexes operated with a *system* of manufacturing that was set up for continuous flow from raw cotton to fabric. That system was most economical when it ran at full capacity. Running only part of the machinery was difficult, since managers could not simply shut down one part of the process for any significant period of time. Machines at one stage of production fed the next stage, and material stacked up quickly if any part of the system was down. To run at reduced capacity, a mill superintendent would have to keep the system in balance, stopping some machines in each department and laying off operatives throughout the complex. It was hard enough to recruit and keep good workers without adding the threat of temporary unemployment. Since a consistent level of surplus waterpower would not be available throughout the

A Harris-Corliss steam engine of the type often used in Lowell as part of hybrid (steam and water) power systems. Courtesy of Slater Mill, Pawtucket, Rhode Island.

year, a new mill that had no supplemental steam power faced high costs for the installation of machinery that would not always be in operation.

In a letter to the president of Locks & Canals in 1859, Francis had predicted that "the result of the extensive use of the surplus power at Lowell will be I think, to run it in connection with steam power."[6] He was prescient in his recognition that coal-fired steam and surplus water would be mutually supporting. Lowell had a locational disadvantage when it came to steam generation. Coastal cities in southern New England could get their coal by ship or barge, a less expensive delivery system than the railroad, on which Lowell's manufacturers depended. It was the availability of surplus water at reasonable cost that made hybrid (water and steam) power systems in Lowell competitive with 100 percent steam systems in New Bedford, Fall River, and Providence.

It is possible that Lowell's industrial expansion would have stalled without surplus water. No additional *permanent* millpowers were sold after 1853. Why would you build textile mills in Lowell that required steam power, when you could go to relatively nearby cities where coal was cheaper and other manufacturing costs were similar? If your mill complex could get surplus water for most of the year, the decision to invest in boilers and steam engines at Lowell made more sense.

As the chief engineer of Locks & Canals, Francis advised the textile corporations on both waterpower and steam power, as well as on the interconnected power transmission systems that allowed their combined use. Eventually, some of Lowell's cotton manufacturers installed enough steam capacity to run all their operations without any waterpower at all, but they continued to use their hydraulic turbines to reduce coal consumption. When floods created extreme backwater or forced the closing of canals, the hybrid mills did not have to halt much, if any, of their machinery. That security was an added benefit of interconnected power systems, which could make use of either steam or water.

Dreams of surplus water had played a role in the formation of the reservoir system in New Hampshire. Although the strongest motivation for buying the lakes was to increase the total number of millpowers that could be supplied year round, it was obvious that those reservoirs could also provide a lot of surplus water. It is no wonder that Francis made close monitoring of lake levels and control of outlet flows a regular part of his job. Within minutes of his sending a telegraphic message to New Hampshire, a loon floating quietly on Squam Lake (Golden Pond, in the Hollywood film) would find his or her elevation dropping slowly but steadily as more water was released to feed the mills in Lowell and Lawrence.

Francis never dominated the Merrimack River, but his engineering projects did affect its seasonal flow patterns, reducing the frequency, duration, and seriousness of low-water conditions. Surplus water fees were a way to profit from natural variations that no engineer, even one as capable as Francis, could eliminate. The fees that he charged also restrained excessive use of water on a canal system with built-in limitations.

The Concord River was another source of waterpower in Lowell. In the relatively short distance of two miles, it offered more drop than Pawtucket Falls, but its average flow was far less, and it lacked the large reservoirs that made the Merrimack so dependable. The Middlesex Manufacturing Company had rights to most of the flow at the first dam above the confluence of the Concord and the Merrimack. In good conditions, the Middlesex could develop almost 200 horsepower from the Concord. That mill complex also received water from the Merrimack River through the Pawtucket Canal and had its own boilers for steam power.

The next dam on the Concord River powered only two mills, but a third dam at a higher elevation fed the 2,300-foot Wamesit Canal. Begun in 1821 by Oliver Whipple, the canal had become a source of energy for seven different mill sites. Unlike Locks & Canals, the Wamesit Power Company did not lease guaranteed mill-powers or sell surplus water. Instead, each user paid for rights to a specified fraction of the water available at any time. This was the most common way of dividing power among multiple users at an American water privilege, but none of the mills on the Wamesit Canal could accurately determine its share of the variable flow.

Much more water was available at the principal drops on the Merrimack, and it was an increasingly valuable resource. The men who ran Locks & Canals and those who operated the Essex Company in Lawrence had a financial incentive to minimize any "waste" of water over their dams after they began to charge for surplus. They could not prevent some spillage, particularly in freshets; but any flow that did not drive a wheel, feed a boiler, or assist in an industrial process (scouring, bleaching, dyeing, printing, etc.) was water for which they could not bill. This was an important factor in their objections to fishways, which took water from the millponds.

Few, if any, anadromous fish were getting by the dams at Lawrence and Lowell during the Civil War. In 1864, officials in New Hampshire complained bitterly to their counterparts in Massachusetts. Commercial fishermen in Lowell and Lawrence were also upset by the failure or blockage of fishways on the dams and the resulting destruction of their livelihood. In hearings held in Massachusetts in 1865, Francis argued that a fishway at Pawtucket Falls would mean less waterpower for the mills in the driest months of the year. At those times, "the water is substantially all used by the several manufacturing establishments to whom the mill privileges have been sold." Providing water for a fishway was less of a problem in the spring, "but it would be at the expense of the mills later in the season."[7]

With the state preparing to pass a law in 1866 that would bring fish migrations back to the Merrimack River, Francis told Storrow at the Essex Company that "under ordinary circumstances I should deem it very objectionable to us, but I think it is as well as we can expect."[8] Both men were also under heavy pressure from the governor and legislators in New Hampshire, where the crucial reservoir system for the river was located.

After the enactment of the legislation, Francis met with the Massachusetts fish commissioners in Lowell to get their ideas about fishway design. He also went with them to interview a man who had fished commercially at North Andover, below the Lawrence Dam, for thirty-six years. Mr. Hardy was still catching shad, but the last salmon he had taken was in 1862. "He had not heard of any being caught in the Merrimack River since." In September, the chief engineer recorded the following action: "Built Fishway at the main Dam at Pawtucket Falls, near the Dracut shore, in accordance with the Act of the Legislature passed at their last session."[9]

Fish-stocking programs and improved fishways produced positive results in the 1870s, with salmon returning to New Hampshire waters in significant numbers. In 1874, the same veteran fisherman, Mr. Hardy, landed "the first salmon caught in the Merrimack River, so far as he knew, since the attempts have been made to restock the river."[10] That salmon's weight suggested that it had been spawned about 1868, after the fishways had been installed at Lowell and Lawrence. Despite the promising benefits of stocking and fishways, overfishing would take a heavy toll of fish populations in the river in the 1880s.

In 1883, Francis was upset by what he considered unfair criticism of his dam and fishway. After reminding a fish commissioner that the dam did conform to the state's requirements, he added that "the usual space at the westerly end of Pawtucket Dam has been kept free of flashboards so far this season, and Salmon, as well as all other fish, have the same rights and privileges as heretofore."[11]

Protecting the rights of the canal company and its corporate owners was one of Francis's principal responsibilities. He knew that water from the Merrimack River was the great renewable asset of Locks & Canals. The "chief of police of water" would not give up any of that water without adequate compensation unless he was forced to do so. When some houses were found to be taking a little canal water for domestic purposes, Francis proposed charging them.

The danger of drinking polluted water was not his main concern in this case, even though he knew that his canals were waste conduits for process water (dyeing, etc.) and human wastes from mills. He would not allow urban sewer connections with the canals, but his corporate customers often treated canals and tailraces as convenient sinks. There was little regard in Lowell for what went into the river, either directly or indirectly.

What really worried Francis was the increasing water needs of the City of Lowell, which had been considering a public water supply system since the 1830s. Providing a trickle of water for a few tenements along the canals was one thing, but letting an entire city use the river for drinking, cooking, washing, fighting fires, and wetting dusty roads was something else entirely. Francis had apparently been willing to consider such a system in 1848, when Locks & Canals stood to benefit from its use for corporate fire protection. At that time, the northern lakes and the new Northern Canal promised much more year-round flow, and the city was talking about leasing the unused power at Hunt's Falls (just downstream of the city center) from the canal company to pump its water. Things changed quickly after Locks & Canals got its own reservoir for fire protection in 1849. By the mid 1850s, many of the corporations were taking as much water as they could get from the canals for their manufacturing.

In 1855, when the Massachusetts legislature first authorized the city to develop a public water supply system, Lowell's major corporations made sure there was an important restriction in the state act: the city could not use water for power. The industrial interests did not have enough political influence to prevent the city from taking water out of the Merrimack River above Pawtucket Dam, but they dampened any hopes for a water-powered pumping station, and they also secured a requirement for compensation if the city did take water controlled by Locks & Canals. Despite this enabling legislation, nothing happened for years, as the Panic of 1857 and then the Civil War intervened.

By the late 1860s, Francis had become a visible spokesman for the corporate opposition to public water in Lowell. The residents of the city had shown a great deal of respect for him since his gate held back the floodwaters in 1852. They knew of his many contributions to local fire protection and appreciated the international acclaim for *Lowell Hydraulic Experiments*. He had served multiple terms as one of their aldermen and represented them in the state legislature in 1869. People listened to his opinions in Lowell, particularly on the subject of water.

Being well read in the literature of public works engineering, Francis recognized that a municipal water supply, if properly managed, would make Lowell less vulnerable to fire and probably to outbreaks of water-borne diseases like cholera, which were already menacing his and other communities in Massachusetts. But, as a representative of the corporations that paid over half the taxes in the city, he was concerned about the cost of the proposed waterworks and a citywide distribution network. And, as the manager of a canal system with rights to all the flow in the Merrimack, he feared that using river water for the city could mean shortages for the mills in dry seasons. In reality, the plan for a municipal system was a much bigger threat to surplus water sales than it was to the delivery of "permanent," contracted millpowers. Francis was a public figure whose integrity was usually above question, but his stiff resistance to Lowell's plans for public water brought a great deal of criticism, and even accusations that he used dubious engineering arguments to further corporate interests.

With fires doing enormous damage to American cities and polluted water increasingly suspected as a major cause of epidemics, this was one fight he could not win. Could his futile, and probably counterproductive, campaign against a municipal water system have been a last ditch effort to gain preferential tax treatment or a larger payment for water the city seemed almost certain to take? Clearly, Francis paid a price in diminished public esteem for his stand against this public expenditure. A widely circulated pamphlet that he wrote in February 1869, while people were preparing to vote on building a water system, seemed to exaggerate the price tag for an urban waterworks and raised the specter of lasting damage to the textile corporations that were the lifeblood of the city's economy.

Even newspapers that usually showed nothing but admiration for Francis ran letters attacking his arguments and his efforts to influence the vote on the "Water Question." One published letter described Francis's pamphlet as the "thunderbolt of corporate wrath" and depicted the views of the chief engineer and his employers as "utterly and entirely selfish." It claimed that the textile manufacturers were "opposed to the introduction of water because they already have it, and because the burden of extending it all through the city will have in part to be borne by the corporations." The writer also showed personal disappointment in Francis: "As a matter of business, we can forgive him, but as a public-spirited, large-hearted, and sagacious citizen, as we all wish to regard him, I am unable to reconcile him to the anomalous position in which he now appears."[12]

A writer in another newspaper also took offense at Francis's pamphlet: "Like any man true to the interests of his employers, the author makes out the worst possible case" and uses "guesswork" to exaggerate the costs. This self-styled "water consumer" was particularly upset by the implied threat of corporate retribution if voters approved the expensive water project. Francis had warned in his pamphlet that "if this money is paid it will come out of somebody, and you may be sure the stockholders are not going to suffer the loss if there is any possible way of putting the burden on some one else."[13]

After weighing the less expensive option of taking water from Beaver Brook (which Locks & Canals did not control), the voters and their elected city officials decided to use the Merrimack River and pay whatever damages were necessary to the canal company. A *Courier* editorial suggested that Francis's pamphlet had been counterproductive. It was "not only inherently weak, but it was a damaging evidence of the weakness of the cause it was designed to defend; and it stirred up such an exciting interest and such a lively opposition, that the friends of water ought to extend a vote of thanks to its author as their most valuable ally."[14]

Francis had not done well in his public relations campaign to stop the water project, but his understanding of hydraulics and his extensive knowledge of waterworks practice in both Europe and America were still regarded as valuable assets for the city. Some of his technical concerns about the pumping and storage plans were valid. Despite his public record of opposition to the project, he was promptly elected to the city's Board of Water Commissioners, which was to oversee the design and construction of Lowell's new municipal water system. In arguing for his election, the editors of *Vox Populi* said that the water question was now settled "and now comes the matter of constructing the waterworks, when we want the best skill and experience that we can obtain. Who is better qualified than Mr. Francis, whose whole life has been spent in an employment qualifying him for the position for which we have suggested his name?"[15]

Lowell built its steam-powered waterworks between 1870 and 1872. A steam engine pumped from the millpond in the Merrimack to an elevated city reservoir, which then supplied water under constant pressure for an expanding network of pipes. The city made a one-time payment of $50,000 to Locks & Canals for taking the company's water above the falls. This was not enough to satisfy the chief engineer or his board of directors, but it helped to offset some of the income they expected to lose from surplus water sales. Having no further need for flow from the Locks & Canals fire mains and concerned about the cleanliness of their water, the city eliminated its annual payment for using them. Although the corporations kept their own fire protection system, they made new one-way connections that could send city water into their pipes in an emergency.

Industry paid most of the costs for the city's new waterworks through taxation and had to give up a little of the river's water, which was important for powering mills in dry months. There were benefits for the companies, however. Improved safety and health were as important to corporate employees as they were to any other residents of Lowell. The *Courier* found "water from the new water works" to be a great improvement: "Those who have been depending on contaminated wells or foul cisterns for their drink but have now the city water, find the change one which is too beneficial to be adequately described. We expect to see a marked diminution in mortality next summer over previous years."[16]

Despite the high hopes of the *Courier*, the increasingly contaminated waters of the Merrimack, without sand filtering, turned out to be better for fire protection than for human ingestion. Waterborne pathogens struck hard at the local population in the 1880s and 1890s.

Francis was able to continue supplying his corporate customers with a great deal of surplus water, applying lengthy prohibitions only during the exceptional droughts of 1880 and 1883. Power from surplus water remained cheaper than steam power (except in some operations requiring process heat) throughout the chief engineer's life. It was in the canal company's interest to set fees that would encourage the use of water. In 1885, the directors of Locks & Canals lowered their minimum fee for surplus water to $4 per millpower per day. One of the major reasons for that action was "the reduced cost of steam power from what it was in 1870 when the present rate of five dollars per mill power, per day, was adopted."[17]

Although every cost saving counted in the highly competitive textile industry, the need for surplus water eventually diminished. In 1888, the directors asked Francis (then retired) to give his opinion on whether they should sell the northern reservoirs, which legislators and industrial interests in New Hampshire were pressuring them to do. He predicted that there would be "a greater irregularity in the supply of water" but that this problem would become less important over time. Francis favored selling the lakes. He argued that waterpower "cannot be relied upon to operate the increased amount of machinery, except to a comparatively small extent during part of the year. The reliance must be on steam power." Mills would need enough steam power to run their textile machines in emergencies, but they would also use waterpower "as far as practicable to reduce the consumption of coal. A large proportion of the water powers in New England are already operated on this basis."[18]

Using waterpower "as far as practicable" did mean avoiding unnecessary losses. Less water was wasted when all the mills started and stopped their machinery at the same time. Every corporation on the canal system was supposed to follow the annual *Time Table of the Lowell Mills*. This printed schedule listed the hours of operation, noting the exact minute that production was to begin and end. When they hoisted their head gates each morning to tap the stored energy in the pond, employees of the canal company anticipated the simultaneous starting time at all the mills. Leakage overnight might have lowered the water in the canals, but Francis made sure that they would be "up to their usual level before the mills start."[19] This early example of "just in time delivery" avoided waste and discouraged corporations from trying to get a jump on their neighbors. When water was scarce, Locks & Canals was unlikely to allow any special use of waterpower or process water at night.

Time was as much a factor in the engineering management of the river basin as it was in the control of the canal system or the supervision of a particular mill. In ordering extra releases of water in New Hampshire, Francis had to anticipate how much he would need and when. He had to decide how long it would take for water to travel more than ninety miles from the lakes to Lowell. That was not easy when mills upstream might be ponding water at night and thus temporarily halting the movement of water in river channels. Sudden freezes could also impede runoff and block streams. In 1858, he told the agent at the Lake Company that he would anticipate "as well as I can what the state of the river would be three days afterwards." That interval was "the time we should be likely to get the benefit."[20]

Francis and his assistants wanted immediate response when they asked for changes in the positions of head or guard gates, but large ones were hard to move. The ten gates in the Pawtucket Gatehouse, for instance, weighed approximately six and one-half tons each. Throughout most of industrial America, workers used gearing or levers to slowly hoist heavy gates by hand. The turbine-driven gates installed at the entrance of the Northern Canal made the Pawtucket Gatehouse a model of labor-saving technology. The operator had to crank open only one butterfly valve to start the turbine. The rest was simply a matter of selecting which gates for the power train to move. In 1852, *Appleton's Dictionary of Machines,* a very influential set of technical volumes, chose Francis's design as its example of "Gates, Guard." According to that source, all ten of the gates could "be lifted at one time, under almost any circumstances, in fifteen minutes." Engineer William Worthen, who considered the gate hoisting machinery a key part of the Northern Canal project, said that the "immense gates . . . worked so readily by water power that they are shut when necessary at noon to pond and retain the otherwise waste water."[21]

The Pawtucket Gatehouse was in a very advantageous position for water-powered operation: it was right at the dam, where a turbine could draw water directly from the millpond, even if the canals had been drained. Although Francis also had a constant source of water at the guard gates on the Pawtucket Canal, there was no fall comparable to the one at the Pawtucket Dam: the millpond and the controlled upper level of the Pawtucket Canal were usually at approximately the same height. Lacking enough drop to run a turbine, the canal company could have relied on a steam engine, but that would have meant keeping a boiler hot or firing it up every time the gates needed adjustment. Instead, Francis had simply paid men to open and close the five guard/sluice gates by hand.

In 1869, the directors approved Francis's plans and estimates "to hoist the gates by steam power." But the chief engineer soon had a better idea. He thought "that it would be better to hoist by means of hydraulic lifts operated by water drawn from the reservoir on Lynde Hill connected with the fire apparatus. . . . The advantages over steam power would be greater certainty in the power being at all times ready for use when wanted; and the greater economy in working."[22] Here was a perfect engineering solution. Francis realized that pistons operating with energy from pressurized water could move very heavy gates. As someone who paid attention to innovative engineering practices in Britain, Francis would have known of Stephenson's and Brunel's success raising enormous bridge sections with "hydraulic lifts" (or jacks) and their extensive use to power the gates in the wet docks at London and Liverpool. His conceptual leap was to link hydraulic lifts to the fire protection system of Locks & Canals. It is not surprising that the board voted unanimously to approve this change of plan.

Francis went back to a turbine when he powered the waste gates on the Great River Wall of the Northern Canal in 1872. Manual operation of those four gates had sufficed since 1847, but it was easy to install a small, scroll-case turbine in one of the adjacent scouring holes used for flushing silt out of the drained canal. The gates emptied the Northern Canal for routine maintenance or repairs and released excess water when necessary.

Power for the guard gates of the Northern and Pawtucket Canals meant that Locks & Canals could respond much more quickly to changing power demands on the system and to variations in the level of the millpond. This reduced the amount of water lost through overflow weirs or waste gates when canals rose above safe levels. It also helped to maintain the heads guaranteed to the corporations in waterpower leases. A smaller number of employees could keep the system in equilibrium, minimizing the occasional slowdowns, which hurt production. Men still struggled to adjust gates and weirs by hand at other places on the system, such as the Swamp Locks and the Moody Street Feeder Gatehouse, but Francis had made a significant improvement in system control with his creative use of turbines and hydraulic cylinders to operate the most important and heaviest gates. In 1877, he added telegraph lines linking the three gatehouses on the Pawtucket Canal to one another and to the Locks & Canals office, another effort to eliminate unnecessary delay and labor (of message delivery).

Regular employees of the canal company did much more than simply follow instructions from an engineer. Men at sites all over the system had to make quick decisions and take action on their own initiative. It took careful observation and much experience to maintain proper flow in the canals without wasting any more water than necessary. When a particular mill complex turned on or off sections of machinery, changed the number of waterwheels in operation, or suffered damage to a critical part of its power transmission system, Locks & Canals had to respond in a hurry. Sometimes the only indication of imbalance in the system was water rising or falling on one level. At the Swamp Locks building, Francis's men might handle such a problem by taking out or putting in place a certain number of the boards that regulated flow through the bays on the crest of the dam separating the upper from the lower level of the Pawtucket Canal. The weir formula that Francis had developed gave an estimate of the flow that would be released by removing a single board, but a worker had to use his judgment in each situation.

Francis was always concerned about seepage through and waste over his main dam at the falls. The crest of the old section remaining from the early 1830s was not perfectly level, due to erosion of loose stone in the riverbed and uneven settlement over the years. Water would begin to spill over the flashboards at low points before the pond had reached its maximum capacity. He complained bitterly about this waste of potential power in an 1869 report to the directors. During dry seasons, he could not afford any loss of water from the pond: "In low stages of the River nearly one half of the supply of water to the mills, is derived from the accumulation in the pond above the dam, during the preceding night." He estimated that the mills at that time were giving up "about 3½ millpowers due to the defects in the present Dam."[23]

Confident that the chief engineer could remedy the problems, the directors finally agreed to rebuild the old section of the dam in 1875. Francis provided a detailed account of the entire project in one of his record books. He had his men rework granite blocks from the old dam for the lower courses. He also ordered a great deal of new stone. The lowest qualified bidder for the principal granite contract was the Cape Ann Granite Company. Moving stone from the north shore of Massachusetts to a region famous for its own granite quarries was like bringing coals to Newcastle, but Francis readily accepted the coastal firm's bid. That meant shipping granite by sea to Boston, then by rail to Lowell, and finally by canal boat to the dam. The Lowell canals were still performing a valuable transportation function.

Cleveland Cheney was in charge of actual construction. He had replaced Paul Hill as "superintendent of outdoor work" in 1863. After working as a carpenter and builder in Lowell, Cheney joined Locks & Canals in 1847 and gained experience on many projects, including the building of the underground Moody Street Feeder in 1848. He apparently did well working for Hill. Francis's son said that "his skill and industry, as a carpenter and a mechanic" put him "at the head of his co-laborers." It did not hurt that he was also "physically very strong." Even as superintendent, he often worked right beside his men on difficult tasks such as clearing ice from the canals "in the dead of night, with the temperature below zero." In warm weather he could be found "in the canals on Sundays and holidays, at times when the water could be drawn off, making necessary repairs, enlargements, etc." When he supervised the recon-

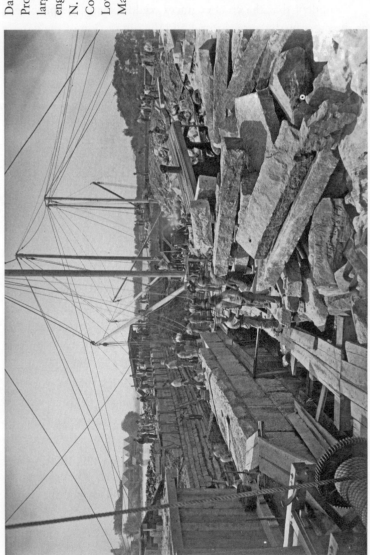

Dam construction in 1875. Projects of this size required a large workforce and careful engineering supervision. N. C. Sanborn, photographer. Courtesy of the Center for Lowell History, University of Massachusetts Lowell.

struction of the dam, "over 400 men were employed, during a large part of the time."[24]

Just before developing a design for the new part of the Pawtucket Dam, Francis completed an investigation of the failure of a reservoir dam on Mill River in Massachusetts for the American Society of Civil Engineers. He made sure that water could not seep under his "gravity dam" and create uplift by hydrostatic pressure. He knew that uplift was the primary cause of the Mill River disaster, which cost 139 lives. The downstream face of Francis's dam was to be made of blocks set directly on the rock, which was cut in steps to secure them. Only in "the deep place" was he forced to depart from this plan. There he had Cheney prepare the irregular bottom by "leveling up with rubble masonry in cement" and then put more than twenty-five feet of ashlar (cut stone) construction on top of it.[25]

The work at the falls included significant alterations of the geological formations that made the site so wild in appearance. Francis cut down three rock projections that had been incorporated in the earlier dam. Once capped with cut granite from Cape Ann, the "fine-hammered" crest was level from the Pawtucket Gatehouse to the northern shore of the Merrimack, a distance that Francis recorded as 1,093.510 feet. The twenty-foot poles used for the measurement could not possibly have met that level of accuracy, but he liked to strive for precision.

Francis put two feet of flashboards back on the dam, cutting new holes in the top stones of the reconstructed section to accept the wrought-iron pins that held up the boards. Leakage was minimized, but nature could still be uncooperative. A terrible drought in August 1883 made it temporarily impossible to satisfy all the leases of water on the canal system. Troubled by this deficiency, Locks & Canals added an additional foot of flashboards to the dam (and made additional "flowage" payments for expanding the millpond). Soon after, the company ordered longer pins made of the new Siemens Martin steel. They would hold the three feet of boards above the stone crest. Francis commented that "the principal effect of the

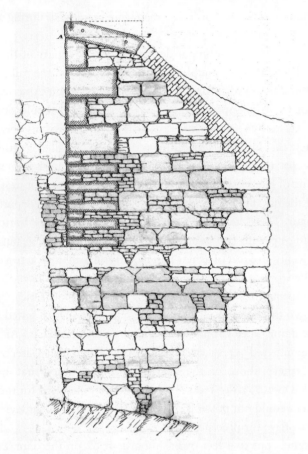

Pawtucket Dam in the "Deep Place," as drawn by James B. Francis in 1875. This cross section in the chief engineer's record book H shows the care he took to securely seat the dam on a particularly difficult part of the river bottom. The millpond is to the right. Courtesy of Lowell National Historical Park.

extra foot of flashboards, is in keeping up the head in the canals and to store a larger supply for use as surplus power."[26]

A continuous wall of flashboards was not on the dam year round. This vertical extension of the crest was only essential when river flow was low. High water washed away flashboards and bent

Bird's-eye view of Lowell in 1876. The population had just reached fifty thousand, and the city was the nation's leading producer of cotton textiles. To the right of the dense cityscape are the dam and the extended rapids of Pawtucket Falls. The Northern Canal runs beside those rapids before turning toward the mills. Howard Bailey and James Hazen, delineators. Courtesy of the Lowell National Historical Park.

Workers replacing flashboards on the dam, ca. 1985. These boards, attached to metal rods, raise the effective height of the dam, increasing the storage capacity of the millpond. Each winter, high water and ice floes tear most of the boards away and bend the rods.

their support pins. This reduced flood hazards by lowering the effective height of the dam. Once water levels had fallen in the late spring or early summer, Locks & Canals employees took on the risky job of retrieving bent pins, inserting straight ones, and raising the flashboards to their established height. According to oral tradition among the employees who still replace the pins and flashboards every year, the canal company gave you a day off if you were accidentally washed over the dam. The National Park Service has restored a small shop at the falls, where company blacksmiths heated and straightened one set of pins every winter.

In 1881, with surplus power still very important, Francis again provided cost estimates for major changes in the canal system. As in 1855, he examined the feasibility of additional feeders between the

Western Canal and the Merrimack Canal. Also included in the new plans were significant enlargements of both the Western and the Pawtucket Canals. The objectives were to eliminate some of the relative disadvantages of mills fed directly or indirectly by the Pawtucket Canal and to increase the amount of surplus power that all corporations could take. The board decided that the potential gains were not worth the high expense for any of these projects.

The Pawtucket Canal remained a source of bitter complaints from mills like the Hamilton and Appleton. Not until 1887 would Locks & Canals begin to rebuild walls on the upper Pawtucket and in the process widen some of the troublesome stretches. After that, the focus shifted to deepening efforts, another way to increase the cross section of this inadequate feeder. Although Hiram Mills, who carried out much of the later work, gave Francis credit for "much improvement in the old canals" during his tenure, the enlargement efforts up to 1885 were limited in scope.[27]

Deepening projects were not restricted to canals. Francis had long considered altering the bed of the river in order to increase the total drop in Lowell. In 1876, he began an ambitious project to cut a channel through Hunt's Falls (at that time, two sequential drops of about six and four feet just downstream of the city), "the effect of which would be to lower the level of the water back of the mills discharging into the Merrimack and Concord Rivers, thereby increasing the fall and power at those mills." Worthen credited Francis's removal of "a portion of Hunt's Falls" and his raising of the flashboards on the rebuilt dam with "increasing the available fall at least fifteen percent, and to that extent the total water power."[28]

Francis's ideas for improving waterpower in Lowell came from diverse sources and often involved the adoption of technology that he had read about in publications, discussed in correspondence with industrialists and other engineers, or seen on one of his many trips in the United States and abroad. His increasing involvement with engineering societies such as the American Society of Civil Engineers, and with educational institutions such as Harvard University and the Massachusetts Institute of Technology, put him in frequent touch with distinguished technical specialists. He also expanded his range of contacts through consulting work, service on commissions, and testimony as an expert witness. He carried on an incredible amount of correspondence on scientific and engineering matters and was a prodigious reader of professional literature.

The directors of Locks & Canals encouraged Francis to travel and to investigate engineering developments that might provide lessons for Lowell. After he had finished the Northern Canal and associated projects, the company directors voted "that Mr. Francis have leave of absence for such a period as he may deem expedient to travel in Europe and that his expenses be defrayed by this company." In 1849, he spent time studying wood preservation treatment in England and visiting the reservoir-fed waterpower system at Greenock, Scotland. In 1851, the directors paid for another trip to Great Britain "to attend the 'World's Fair.' " That Crystal Palace exposition, in which American achievements in manufacturing earned high praise, was the first great fair to showcase the technical achievements of the world. Paul Hill, who was left in charge during at least one of the chief engineer's extended absences, said that Francis went overseas "at least a half dozen times, with his eyes wide open, and his mind on the alert concerning the latest ideas in the progress of mankind."[29] He also traveled widely in the United States and Canada.

A complete set of letters from a European tour in 1879 reveals much about Francis and his insatiable thirst for engineering knowledge. On arrival in Cherbourg, he started looking for things "applicable to Lowell." This included not only material evidence and observations of work in progress but also any useful publications that he could acquire for the Locks & Canals library. His engineering sketches complement descriptions of useful equipment, for example, a floating boom of pine logs, which might have blocked flotsam at the entrance of the Pawtucket Canal but was not as good as "the one Cheney made last fall" for that site.[30]

Francis had arranged ahead of time to see things that would not be on the usual visitor's itinerary. He benefited from guided tours of factories, workshops, and construction sites. Much of his activity was open gathering of information, but some of it was thinly disguised industrial espionage. The chief engineer was following in a well-documented Lowell tradition. The man for whom the city was named had freely borrowed British weaving technology and also picked up the idea for an automatic waterwheel governor. Francis's father-in-law, George Brownell, was amazingly candid about his own spying for Locks & Canals in 1839. As the superintendent of the canal company's machine shop, he visited dozens of British firms that made machinery. The directors sent "Mr. Brownell to England at the expense of the company for the purpose of obtaining information on the subject of machinery & the manufacture of the same."[31]

It is not surprising that Francis showed particular interest in the hydraulic engineering that was his specialty. He wrote that he wanted to examine more of the "great hydraulic works" when he got to Holland. "I want to see some in progress, it is far more instructive than the finished work." In Zurich, there was "not much fall, & of course not much power. Most of the mills were apparently driven by steam, and they were "all small compared with our Lowell mills."[32]

More noteworthy was the waterpower at Bellegarde. He sent home a special report for Locks & Canals and the Essex Company about that French site on the Rhone River. It had enough head and flow to rival Lowell or Lawrence, but Francis criticized its slow development and economic difficulties during an 1881 address to the American Society of Civil Engineers. In general he seemed unimpressed with European waterpower sites because few were developed in what he called "a systematic manner from their inception."

In his address, Francis provided a full description of the "usual process of developing a large water-power" in America. First, investors formed a company and bought enough land for a town "to accommodate the population which is sure to gather around an improved water-power." Then workers built the dam and necessary canals. The company granted "mill sites with accompanying rights

Trees planted along the inland section of the Northern Canal, looking southeast toward the stack of the Tremont Mill. The crooked tree on the right is a hackberry moved by James B. Francis to the head of an existing row of elms. Courtesy of the Lowell Historical Society.

to the use of the water . . . , usually by perpetual leases subject to annual rents." Francis said that this method of development was "distinctively an American idea." He considered Bellegarde to be the only example of this approach in Europe, "but from the great outlay incurred in acquiring the titles to the property, and other difficulties, it has not been a financial success."[33]

Francis was always curious about useful technology wherever he found it, but on his 1879 trip he was not traveling simply to acquire practical information. He was also a sightseer, drawn to natural wonders and cultural attractions. Mrs. Francis, who traveled with him, shared his interests in the gardens of southern Europe and encouraged him to attend the classical theater in Paris, where the engineer who had taught himself to read French technical literature with ease discovered that he could not follow the dialog. He said that all he understood was "excusez moi." They left after the second act.[34]

Lowell was never far from his mind when he was away. While Francis was admiring botanical specimens and landscaping in Italy, he was also thinking about planting trees in his adopted home. In one of the letters, he discussed setting out poplar trees beside the Merrimack River. He wanted them "boxed to keep cattle, etc. off." On the back of the envelope is a sketch map with a notation in his hand: "line for trees to be planted."[35] He was making sure that the layout would be exactly as he planned it.

When Francis returned to Lowell, he continued to work on the beautification of company property, including the very popular "Canal Walk" along the Northern Canal. His orderly spacing of trees along the canal is apparent in a number of historical photographs and on a map published in 1850. Unfortunately, none of these graceful elms has survived the ravages of time, the impact of tenement construction, or the attacks of Dutch elm disease. Also sadly missing from the present landscape is a tree of considerable renown that Francis saved from destruction in 1875.

This large hackberry grew naturally near the Northern Canal

but was marked for cutting because it encroached on a road. Francis had noticed the exceptional quality of the tree. Instead of allowing the sawyers to proceed, he used heroic procedures to reposition the hackberry at the head of a row planted on the canal bank. Company employees dug a trench twenty-one feet in diameter around the tree, waited till cold weather, then hoisted and moved the entire tree with its root system safely encased in a disk of frozen ground. A reporter commented that "when Mr. Francis wants to have a thing done, he is in the habit of finding a way to do it." He added that this successful effort deserved "the acknowledgments of every lover of nature."[36]

Francis saved this noble hackberry, but sometimes the imperative for industrial expansion, urban development, and rail access in the city forced him to compromise. He could not protect every link in the chain of greenery that complemented the canals. Over time he lost much of the landscaping along the Merrimack Canal by the Lowell Machine Shop and at the Shattuck Mall.

It was no longer easy for Francis to stroll the green edges of his canal system. His physical capabilities were growing more limited, but his appreciation of well-designed landscapes and his intellectual energy were as high as ever. The stature that he had achieved as an accomplished engineer brought increasing requests for his professional opinions. His study of European and American technical developments and his level of correspondence did not fall off significantly with age, but he was spending more time away from the Locks & Canals office by the late 1870s. Francis had always given considerable authority to the able engineers who worked for him, including men like Joseph Frizell and Hiram Mills, who went on to build their own reputations as authorities on hydraulic engineering. Now his assistant was the son that he had raised to be an engineer.

James Francis, like his better-known father, James B., became a very capable engineer without any formal schooling in the subject. Born in Lowell, he went to public school only until the age of twelve and then received private instruction for eight more years. James did not pursue a college degree but instead went into the Lowell Machine Shop as an apprentice. He gained practical experience on the shop floor until the Civil War began. Soon after enlisting in 1861, he was promoted to second lieutenant. Although wounded at Antietam, he went on to compile an impressive combat record and attained the rank of lieutenant colonel. Following his discharge in 1865, he worked for several years on the Hoosac Tunnel before joining his father at Locks & Canals, where he was in charge of water measurement and other engineering activities. Worthen said that James B. could devote less "attention to the duties of his office" after the directors "promoted his son to the office of his assistant."[37]

The directors readily granted the leave for Francis's trip in 1879 and appointed his son acting agent. Their high regard for Francis as a person and their gratitude for his exemplary service to the company are obvious in the minutes of their meetings. While he was away, they had a portrait made for the Middlesex Mechanics Association, a group that Francis had long supported, "that it may be placed in their Hall beside the portraits of those gentlemen who we regard as eminently the Founders of the City of Lowell, in order that for all time to come it may perpetuate the likeness and memory of Mr. Francis, whose skill and ability had done so much to set forward the prosperity, not only of the manufacturing interests, but the general welfare of the City."[38]

After his return from Europe, most of the chief engineer's work in Lowell was routine. With no great construction projects under way and with his son handling much of the work in the office, Francis accepted his greatest professional honor and the duties that went with it. In November 1880, he became the president of the American Society of Civil Engineers, holding that influential position until January 1882. He had been active in that national society since being elected to membership at its first meeting in 1852, and he had contributed many important papers and discussion comments in the intervening years. Although he had served as the president of the Boston Society of Civil Engineers in 1874, the national presidency was a much greater accomplishment and proof of the respect he had earned from his colleagues.

Worthen said that Francis had a "reputation as the best and most reliable hydraulic expert . . . and his advice was sought throughout the entire country." Mills noted "his wide practice throughout the country in hydraulic work, in which department he has been regarded as the head of his profession in this country." He gave advice on "fifty water powers in nine states and two provinces" as well as "water works in eighteen cities of five states and one province."[39] Successful participation in great works of nineteenth-century hydraulic engineering brought professional prestige and public acclaim in America.

A partial list of Francis's many outside consulting projects, some of which involved litigation, includes improvement of New York's Croton water system; evaluation of the effects of backwater on Boston's Back Bay tidal power; measurement of energy consumption by machinery, belting, and shafting in Atlantic Cotton Mill #2 at Lawrence; investigation of the collapse of the Pemberton Mill in the same city and subsequent publication on cast-iron columns; tests of the Swain and the Humphrey patented waterwheels; gauging of flow for the Cohoes Company in New York; conservation of St. Anthony's Falls at Minneapolis; examination of the catastrophic failure of reservoir dams on the Mill River in Massachusetts and on South Fork Creek (Johnstown flood) in Pennsylvania; and planning for reconstruction of the great Holyoke dam in Massachusetts. He also studied the Augusta Canal System in Georgia; a cofferdam at Turner's Falls, a dike at Provincetown, water diversion on Stony Brook in Waltham and Weston, and foundations for Trinity Church tower in Boston, all in Massachusetts; and irrigation and hydraulic mining in California.

Sometimes a brief statement from Francis was enough to prompt a course of action or settle a dispute. When Patrick Walsh was arguing with Charles Estes about whether waterpower fees in Augusta, Georgia, were comparatively cheap in 1874, he knew whose opinion would count. As he reported in the local newspaper, "I concluded to telegraph *The Authority* upon water power in New England, Mr. James B. Francis in Lowell. I presume Mr. Estes has heard of the gentleman."[40]

In 1882, the directors voted to give Francis four months of paid vacation time per year. He began to take more frequent leaves after that. Both Francis and his employers knew that he was approaching the fiftieth anniversary of his arrival in Lowell on November 22, 1834. He tendered his resignation two days before that milestone, but no one was surprised to see him step down. The directors already had a letter ready to send him to mark the anniversary. The letter declared: "To the eminent ability and wisdom which have distinguished your administration the marked success of the Lowell manufacturers has been due."[41]

Once they had received official notice of Francis's intention to retire at the end of 1884, the directors made him the company's consulting engineer as of January 1, 1885. He was then sixty-nine years old. Simultaneously they promoted Colonel Francis from assistant engineer to his father's former position as both agent and chief engineer. James B. received a consulting payment (or retirement compensation) of $7,000 per year and was also allowed to stay in his impressive company-owned house as long as he wanted, with no rent and with Locks & Canals paying property taxes and making any necessary repairs. By contrast, his replacement got a salary of only $4,000 per year, another $1,000 for house rent, and use of a company horse and carriage.

In his usual meticulous way, Francis inventoried everything in the Locks & Canals offices on the day he turned over the reins to his son. His detailed list is like a tour of the place in which he worked and the material culture that surrounded him. The company library already had its own detailed catalog; cabinets held row after row of notebooks, journals, daybooks, waste books, scrapbooks, and other records; his "Francis Files" on hundreds of subjects were well organized for future reference; and engineering drawings were shelved and carefully numbered for retrieval. He would remain active in the affairs of the Proprietors of Locks and

Canals and would provide valuable engineering and managerial advice for years to come, but he was no longer in charge. Francis knew that the Merrimack would keep flowing without him and that the most renowned direct-drive waterpower system of the world was in good hands.

Postscript

Lowell is the quintessential city at the falls. Creating any permanent human settlement is a technological as well as a social process, which necessarily involves interaction with and impacts on the natural environment. The physical form of most communities is the result of a great number of private and public decisions. Cities at substantial drops in rivers are no different, but topographical and hydrological constraints limit the range of options at these sensitive, sometimes dangerous locations. Lowell became a thriving industrial center because of the waterpower at Pawtucket Falls. In the city today, one is still keenly aware of the presence of the Merrimack River and of the network of canals that fans across the urban landscape.

The canal system built by the Proprietors of Locks and Canals had few rivals for size or complexity. Its five miles of open channels, underground conduits, gatehouses, guard gates, dams, sluices, and spillways were the product of multiple planning episodes, significant additions, and incremental improvements. The early corporate decisions that transformed the old Pawtucket Canal into a feeder for an expansive two-level system were expedient, but they created lasting problems for engineers who had to manage the distribution of water to ten mill complexes and a large machine shop. What has been done before affects what can be done later.

Planners of other nineteenth-century industrial communities, including those built by some of the investors we call the Boston

Associates, saw Lowell as a model but never tried to replicate its exact form. It was a daring experiment in many ways, including its extensive application of direct-drive waterpower for textile production. The city had enormous influence because of its economic success and the confidence it inspired in imitators. It demonstrated that a busy manufacturing center could be clean and attractive, with tree-shaded corridors and handsome buildings. It suggested that America's abundant waterpower was a natural resource that engineering talent and capital investment could harness on a large scale. The canal system in Lowell might have been overly complicated and the urban plan too focused on the optimal siting of mills and workers' housing, but the astounding corporate profits and rapid population growth were undeniably impressive. Many places vied to be "the Lowell" of their region.

For almost half a century, the awesome responsibility for operating and improving Lowell's canal system fell on one man: James B. Francis. He was arguably the finest engineer in nineteenth-century America. His technical versatility and encyclopedic knowledge seem astounding in this modern age of narrow specialization. From 1845 to 1885, he was both the agent and chief engineer of Locks & Canals, a company that was then wholly owned by the corporations that used its water.

Francis sought simple solutions to even the most serious problems. He could be innovative in design and progressive in his promotion of new technology, but the things he proposed had to work well. They had to be better than the things they replaced, in the way that turbines were clearly superior to breast wheels. His emphasis was on functionality and durability. The mighty Merrimack has tried its best, as all rivers do, to remove the obstructions in its path to the sea, but so far it has done little damage to the material record of Francis's engineering in Lowell. His colleague William Worthen said that he did "not know of anything of his that failed from weakness of construction."[1]

Lowell's citizens remember Francis because of his "Great Gate," which protected the city in two catastrophic floods. His concern for the landscaping of corporate property is less well known but also worthy of recognition. Historians of technology study his applications of scientific engineering to the problems of flow measurement and turbine design. Although his own inward-flow turbines were not highly successful, he was a true pioneer, inspiring others to experiment with and perfect that wheel form. The rigorous methods of analysis and testing that he and Uriah Boyden used in Lowell set high standards for the emerging profession of hydraulic engineering. His publication of *Lowell Hydraulic Experiments* in 1855 earned worldwide praise for its author and his path-breaking research. He also had, in Worthen's words, "wonderful administrative and executive ability."[2] Francis was a superb manager whose technical skill, integrity, and judgment earned the respect of corporate executives and helped avoid costly conflicts over water.

The principal goals of Locks & Canals were greater efficiencies in water delivery and power generation. The enemy was unnecessary waste of energy. In its zeal to conserve Merrimack River water, the principal source of energy for the mills, the canal company sometimes resisted the operation of fishways designed to move wild populations of salmon, shad, and alewives past its dam. It also fought but could not prevent the creation of a municipal water system that would withdraw water from its millpond.

Frequent measurement, constant observation, and quick response were required to keep the complex canal system in dynamic equilibrium. And control over water extended far beyond the city limits. Entrepreneurial audacity (which we might see today as arrogance) converted a hundred square miles of New Hampshire lakes into giant reservoirs, altering the runoff patterns of the entire Merrimack drainage basin, increasing the permanent millpowers at Lowell and Lawrence, and assuring substantial surplus flow most of the year.

We should not underestimate the role of waterpower in driving America's industrialization. The energy from falling water was as valuable for manufacturing in the nineteenth century as the energy from petroleum is today. Although Locks & Canals leased no additional millpowers after 1853, the commodification of water in Lowell intensified. Closely regulated sale of surplus water helps to explain the continuing expansion of textile production on the canal system. Coal-fired steam power became increasingly important in the city after the Civil War, but primarily as a supplement, not as a replacement, for less expensive waterpower. Most mills with hybrid power systems used surplus water when it was available in order to reduce the cost of coal delivered by rail. In that way they could compete with manufacturers in coastal cities like Fall River and New Bedford, which had relatively little or no waterpower but good links to the maritime coal trade. Statistics on the total power of installed engines in Lowell are deceptive; much of the steam capacity was usually in reserve, operating primarily in dry seasons, when surplus water was scarce, or in freshets, when backwater reduced the head at riverside mills.

After Francis retired in 1885, the canal system remained a critical part of the industrial infrastructure of Lowell. The sale of the Lake Company (and its northern reservoirs) in 1889 meant that Locks & Canals no longer had control over releases of water in droughts, but mills still used a great deal of surplus water, and the canal company still measured and charged for it. Fellow engineer Joseph Frizell said of Francis in 1894 that "it was no small tribute to his judgment that the methods of water measurement adopted by him at Lowell more than 50 years ago are still in use there, being probably the only mechanical art or process practiced in that city which has not undergone great and repeated changes since that time."[3]

Francis advised Locks & Canals and carried on an active consulting practice until just before his death in 1892. Frizell wrote that during his last days, "his countenance wore an aspect of majestic repose, as of one conscious of having endeavored well and ready to meet the judgment of mankind with equal indifference to praise or censure." A memoir by professional colleagues remarked on his "wide reputation" and said that engineers "have good reason to hold in high regard the memory of this man, upon whose like we shall not look again."[4]

Both Francis's career and the age of direct-drive waterpower came to an end at roughly the same time. He had foreseen not only the increasing reliability and falling cost of steam power but also the potential for electric motor drive. Electrical power, generated by either steam or water, would eventually change the way factories operated and make elaborate canal systems like the one in Lowell seem like anachronisms. A hydroelectric station at or near a waterfall, connected with transmission wires, could deliver power efficiently to mills far from the hazardous banks of a river. In the twentieth century, electric motors, used first on sections of line shafts and later on individual machines, transformed shop floors.

The electrical age came gradually to Lowell's textile mills and machine shops. Southern cotton textile production, which had "mushroomed" since 1880, made some corporate managers hesitant to invest heavily in new technology for the city's aging mills.[5] Locks & Canals continued to use its canal system to provide water to the mill complexes, all of which installed new hydroelectric units or linked generators to some of their existing turbines by 1918. Despite the general adoption of electric motors in the city's mills, the Lawrence Manufacturing Company still used nothing but "mechanical drive" in 1927.[6] The engineers and directors at Locks & Canals had considered building a central power station near Pawtucket Falls as early as 1894, but they delayed until the sharply declining fortunes of local industry in the 1920s and the Great Depression forced Chief Engineer Arthur Safford to shelve the plan. Only three of the original textile corporations were still manufacturing at their own sites by the mid 1930s, and the last of them ceased operations in the mid 1950s. Hydroelectric power, however, remained a saleable commodity.[7]

Locks & Canals kept supplying water to hydroelectric facilities in Lowell's remaining mill yards. When the author helped to survey the canal system in the 1970s for the Historic American Engineering

Record, Boott Mills had survived as a "sister corporation" to Locks & Canals and had become the owner of a number of mill properties, including five operating power plants.[8] Mel Lezberg, who managed both corporations, was in charge of the canal system. He also rented industrial space; sold hydroelectric power, steam, fire suppression, and process water to his tenants; and had a two-way arrangement with the regional utility, Massachusetts Electric Company, which filled electrical needs when waterpower was insufficient and bought surplus energy when it was available. By then, Lowell's waterside elms were gone, and no one had put floating tubes in the canals for flow measurement since the early twentieth century. But the old system still operated well, with a great deal of historic equipment in regular use. You could visit the Swamp Locks control center at any time and find someone "working the water"[9]: monitoring gate settings, making decisions about which turbines should run and how hard, keeping the pond near the top of the flashboards, minimizing waste, and maintaining balance on both levels of the system.

The remarkable survival of the historic canal system was a key factor in the successful campaign to create the Lowell Heritage State Park in 1974 and both the Lowell National Historical Park and the Lowell Historic Preservation Commission in 1978. Public and private efforts by the city, the state, and the federal government, as well as popular support from Lowell's citizenry, have led to the conservation of this cultural and engineering landmark and the restoration or re-creation of key features. Few places have put as much effort into the preservation and interpretation of industrial heritage, or done it as well, as Lowell.

Today you can walk on the paths of Lowell's expanding Canalway, visiting structures mentioned in this book and gaining a sense of the original landscaping from canal-side greenway development by the National Park Service. You can also take guided boat tours from the Swamp Locks, through the Guard Locks, and into the millpond above the Pawtucket Dam and Gatehouse. The canal system is still producing valuable power on the two levels at four different locations: in the former Hamilton, Massachusetts, and Boott mill yards and, most importantly, at the Eldred L. Field Hydroelectric Plant, commissioned in 1985 at the bend in the Northern Canal. You may feel the current in the Pawtucket Canal as you head upstream on your tour, but that will ease as you slip into the stone chamber of the Guard Locks and under the heavy timbers of the Great Gate. If you are in the vessel appropriately named the James B. Francis, that evocative experience should have special meaning.

Notes

··

The notes published in this volume are only for quotations in the text. A more complete set of notes is available online at two institutions, the American Textile History Museum (Chace Catalog) and the Center for Lowell History of the University of Massachusetts Lowell. The following websites provide access to these online notes for *Waterpower in Lowell:*

http://www.athm.org/online_catalogue.htm.
http://library.uml.edu/clh/mo.htm

Works frequently cited in the notes have been identified by the following abbreviations:

CORHA	*Contributions of the Old Residents' Historical Association,* Lowell, Massachusetts
JBF	James B. Francis
LHE	James B. Francis, *Lowell Hydraulic Experiments*
LNHP	Lowell National Historical Park
PL&C	The Proprietors of the Locks and Canals on Merrimack River
PL&C-Baker	Proprietors of Locks and Canals Collection, Baker Library Historical Collections, Harvard Business School
PL&C Directors	Records of the directors and proprietors, Proprietors of Locks and Canals Collection, Lowell National Historical Park
PL&C-LNHP	Proprietors of Locks and Canals Collection, Lowell National Historical Park

Introduction

1. *Hand-Book for the Visiter to Lowell* (Lowell, MA, 1848), 7.
2. James Montgomery, *A Practical Detail of the Cotton Manufacture of the United States of America* (Glasgow, 1840), 16, 162.
3. *Hand-Book*, 9, 34.
4. Adams and Clay quoted in Thomas W. Lewis, *Zanesville and Muskingum County, Ohio*, 3 vols. (Chicago, 1927), 1: 361–65.
5. Nathan Appleton, Journal of 1810, entry for Oct. 16, 1810, Nathan Appleton Papers, Massachusetts Historical Society.

One: Harnessing the Merrimack River

1. Henry A. Miles, *Lowell As It Was, and As It Is* (Lowell, MA, 1845), 14.
2. JBF memo on Pawtucket Canal, Mar. 1876, DB-8, 393, PL&C-Baker.
3. Duke de Rochefoucauld, quoted by Bland Simpson in *The Sierra Club Wetlands Reader*, ed. Sam Wilson and Tom Moritz (San Francisco, 1996), 29.
4. Wilkes Allen, *History of Chelmsford* (Haverhill, MA, 1820), 70.
5. W. R. Bagnall, "Paul Moody," *CORHA* 3 (1884–87): 67–68.
6. F. C. Lowell to Peter Remsden and Co., Dec. 31, 1814, Lowell Collection, Letterbook, vol. 4, Massachusetts Historical Society.

Two: Building a City at the Falls, 1821–1836

1. Nathan Appleton, "Introduction of the Power Loom and Origin of Lowell," in *Development of the American Cotton Textile Industry*, ed. Georges Rogers Taylor (New York, 1969), 17–18.
2. Ibid.
3. Edward Everett, "Fourth of July at Lowell (1830)," in *The Philosophy of Manufactures: Early Debates over Industrialization in the United States*, ed. Michael Folsom and Steven Lubar (Cambridge, MA, 1982), 285.
4. John A. Lowell, "Patrick T. Jackson," *COHRA* 1 (1874–79): 201.
5. Appleton, "Introduction of the Power Loom," 18.
6. Ibid., 18–19.

7. Ibid.

8. PL&C Directors, Nov. 14, Dec. 26, 1821, and Jan. 8, 1822.

9. Appleton, "Introduction of the Power Loom," 24.

10. Daybook 14, Jan. 2, 1881, PL&C-LNHP.

11. John O. Green, "Autobiography," *CORHA* 3 (1884–87): 233–35.

12. Ibid.

13. Ibid.

14. Boott's diary quoted in William R Bagnall, "Sketches of Manufacturing and Textile Establishments . . . ," ed. Victor Clark (microfiche), 2157–62.

15. David Moody, quoted in James Russell, "Biography of John Dummer," *CORHA* 2 (1880–83): 97; William Worthen comment on Robert Allison, "The Old and the New," *Transactions of the American Society of Mechanical Engineers* 16 (1894–95): 747–49.

16. Nathan Appleton, MS. Dated Apr. 15, and surely written in 1823, Nathan Appleton Papers, Massachusetts Historical Society.

17. Ibid.

18. Worthen comment, "The Old and the New," 747.

19. Merrimack Manufacturing Company Directors' Records, May 19, 1824. Baker Library Historical Collections, Harvard Business School.

20. Ibid., Oct. 19, 1824.

21. Ibid., Nov. 22, 1824.

22. P. T. Jackson to Rufus King, Oct. 6, 1824, Lee Papers, "B" Letters of P.T.J., Massachusetts Historical Society.

23. PL&C Directors, Mar. 17, 1826.

24. See *Chelmsford Phoenix*, Oct. 7, 1825, to Feb. 24, 1826; *Merrimack Journal*, Mar. 24, 1826, to Jan. 5, 1827; and *Lowell Journal*, Mar. 2, 1827; Records H, 112, PL&C-LNHP.

25. Henry David Thoreau, *A Week on the Concord and Merrimack Rivers* (Mineola, NY, 2001), 153.

26. Tim O'Rous, "The American Venice," *Lowell Citizen,* Feb. 8, 1884.

27. *Essex Gazette* (Salem, MA), Aug. 12, 1825, quoted in John O. Green, "Historical Reminiscences," *Proceedings in the City of Lowell at the Semi-Centennial Celebration of the Incorporation of the Town of Lowell*, Mar. 1, 1876 (Lowell, MA, 1876), x.

28. Quoted in Christopher Roberts, *The Middlesex Canal, 1793–1860* (Cambridge, MA, 1938), 153.

29. Ithamar Beard, "Practical Observations on the Power Expended in Driving the Machinery of a Cotton Manufactory at Lowell," *Journal of the Franklin Institute* 11 (Jan. 1833): 6–15.

30. Robert Israel to Lewis Waln, Aug. 9, 1831, in Waln Collection, Historical Society of Pennsylvania. Copied by Steven Lubar and Michael Folsom.

31. Andrew Ure, *The Philosophy of Manufactures* (London, 1835), 13.

32. James B. Francis, autobiographical sketch, in PL&C-LNHP, gift of Samuel Francis.

33. JBF Records H, 120, PL&C-LNHP.

34. Desmond FitzGerald, Joseph P. Davis, and John R. Freeman, "James Bicheno Francis: A Memoir," *Journal of the Association of Engineering Societies* 13 (Jan. 1894): 1–3.

Three: Expanding the Waterpower, 1836–1847

1. P. T. Jackson to Directors, Sept. 13, 1839, A1, #7, PL&C-Baker.

2. Ibid.

3. Ibid.

4. "Report of the Committee on the Subject of a New Canal"(Boston, 1840), 4–5.

5. Ibid., 11.

6. Ibid., 12–13.

7. Ibid., 14–15, 20.

8. PL&C Directors, Jan. 9, 1841.

9. *Lowell Courier,* Apr. 28, 1840.

10. Ibid.

11. George Gibb, *The Saco-Lowell Shops* (Cambridge, MA, 1950), 98–101.

12. Ibid., 102–3.

13. JBF to Benjamin Saunders, Oct. 21, 1864, DB-3, 568, PL&C-Baker.

14. JBF to John Morse, Dec. 6, 1845, A34, #177, PL&C-Baker.

15. William Worthen, "Life and Works of James B. Francis," in *CORHA* 5 (1894): 232.

16. M, "Shade Trees," *Lowell Offering,* ser.2, vol. 1 (1841): 233.

17. J.L.B. [Josephine Baker]," The Factory Girl," *Lowell Offering,* ser. 2, vol. 5 (1845): 274.

18. Lucy Larcom, *A New England Girlhood* (1889; Gloucester, MA, 1973), 163; Lucy Larcom to Charles Cowley, Feb. 28, 1876, in *Proceedings in the City of Lowell . . .* (Lowell, 1876), 93–95; Worthen, "Life and Works of James B. Francis," 237.

19. John Morse to JBF, Sept. 27, 1845, DA-2A, 32, PL&C-Baker.

20. Deed, PL&C to Benjamin F. French and George H. Carleton, Mar. 20, 1844. Copy at Middlesex North Registry of Deeds (Lowell), Book 46, p. 138.

21. Ibid.

22. John Greenleaf Whittier, *The Stranger in Lowell* (Boston, 1845), 90–94.

23. *The Protest* (Lowell), Nov. 25, 1848.

24. Israel of Old, *Easy Catechism for Elastic Consciences* (Lowell, 1847), 11–13.

25. Whittier, *The Stranger in Lowell,* 92–93.

26. PL&C Directors, Dec. 4, 1849; JBF, Records A, 181, Oct. 2, 1853, PL&C-LNHP.

27. Deed quoted in Shepley, Bulfinch, Richardson, and Abbott, "Suffolk Mfg. Co.," 1979, 6, LNHP.

28. Elisha Huntington, "Address . . . " (Lowell, 1845), 15–16.

29. Samuel Lawrence to Charles Hovey, Feb. 8, 1875, "Three Letters of Samuel Lawrence, Esq.," *CORHA* 1 (1874–79): 288–89.

30. Merton Sealts Jr., ed., *The Journals and Miscellaneous Notebooks of Ralph Waldo Emerson* (Cambridge, MA, 1973), 10: 102.

31. Henry David Thoreau, *A Week on the Concord and Merrimack Rivers* (Mineola, NY, 2001), 53.

32. D. Spencer Gilman to Moses Jr., Mar. 15, 1846, D. Spencer Gilman Letters in J. I. Little MSS, Center for Lowell History, University of Massachusetts Lowell.

33. JBF Wastebook, no. 5, Jan. 19, 1848, PL&C-LNHP; Gilman to "Canadian friends," Aug. 22, 1845, D. Spencer Gilman Letters.

34. PL&C Directors, Sept. 15, 1846.

35. Samuel K. Hutchinson, "Northern Canal, 1846–1847," in small notebooks, PL&C-LNHP.

36. "Pay Roll of the Mechanics and Laborers . . . ," R-2, PL&C-Baker.

37. Gilman to Moses Jr., July 31, 1846, D. Spencer Gilman Letters.

38. *Lowell Advertiser,* July 21 and 23, 1846; *Lowell Courier,* July 26, 1846, and Oct. 27, 1847.

39. *Lowell Courier,* July 9, 1847.

40. Ledger HB-2; T. L. Lawson to JBF, Dec. 18, 1846, A34, #175, both PL&C-Baker.

41. Paul Hill, "Personal Reminiscences of Lowell, Fifty Years Ago," *CORHA* 5 (1894): 281–82.

42. Ibid.

43. July 7, 1847, NH-2, PL&C-Baker.

44. JBF to John Morse, June 29, 1847, A35, #189, PL&C-Baker; "Paul Hill . . . ," *Lowell Courier-Citizen,* Oct. 31, 1940.

45. *Lowell Courier,* June 24, 1847.

46. Joseph P. Frizell, "Reminiscences of James B. Francis," *Engineering News,* July 12, 1894, 28–30.

47. "The Northern Canal," *Lowell Courier,* Jan. 1, 1848.

48. Hill, "Reminiscences," 281.

Four: Testing the Waters: Scientific Engineering in Lowell

1. *LHE* (1855), xi.

2. Hiram Mills, "James Bicheno Francis," for the Massachusetts Institute of Technology (Cambridge, MA, Dec. 14, 1892), 5.

3. PL&C Directors, Sept. 12, 1840, Sept. 20, 1842, and Oct. 14, 1842.

4. *LHE* (1868), 148.

5. Desmond FitzGerald, Joseph P. Davis, and John R. Freeman, "James Bicheno Francis: A Memoir," *Journal of the Association of Engineering Societies* 13 (Jan. 1894): 3–4.

6. JBF Records A, 125–27, PL&C-LNHP.

7. Mary Blewett, ed., *Caught between Two Worlds: The Diary of a Lowell Mill Girl, Susan Brown of Epsom, New Hampshire* (Lowell, MA, 1984), 33–34.

8. *LHE* (1855), 2.

9. Joseph P. Frizell, "Reminiscences of James B. Francis," *Engineering News,* July 12, 1894, 29.

10. R. H. Thurston, "The Systematic Testing of Turbine Water-Wheels in the United States," *Transactions of the American Society of Mechanical Engineers* 8 (1886–87): 364.

11. *LHE* (1855), 2–6.

12. James Russell, "Biography of John Dummer," *CORHA* 2 (1880–83): 101.

13. JBF, "Experiments and Translations," 73, PL&C-LNHP.

14. JBF memo dated Apr. 15, 1848, quoted in Arthur Safford and Edward Hamilton, "The American Mixed-Flow Turbine and Its Setting," *Transactions of the American Society of Civil Engineers* 85 (1922): 1345–46.

15. Ibid.

16. Uriah Boyden, "Paper on Turbines," PL&C General Files, #1076.3, 4A, 22, PL&C-LNHP.

17. Ibid., 18.

18. *LHE* (1855), 6–7.

19. Ibid.

20. JBF to Boyden, Feb. 1, 1849, JBF, "Experiments and Translations," 185–86.

21. *LHE* (1855), 39.

22. *LHE* (edition abridged for use at MIT, 1885), 12.

23. JBF to Luigi D'Auria, Feb. 25, 1878, A-16, #76, PL&C-Baker.

24. Clemens Herschel, quoted in Hunter Rouse and Simon Ince, *History of Hydraulics* (Iowa City, IA, 1957), 189.

25. Charles Storrow, *A Treatise on Waterworks for Carrying and Distributing Supplies of Water* (Boston, 1835), 3–4.

26. Quotation attributed to Sir Cyril Hinshelwood. See George E. Smith, "Newtonian Style in Book II of the Principia," in *Isaac Newton's Natural Philosophy*, ed. Jed Z. Buchwald and I. Bernard Cohen (Cambridge, MA, 2001), 286–87.

27. *LHE* (abridged for MIT, 1885), 20.

28. *LHE* (1883), xi.

29. Ibid., 118–19, 133–35.

30. William E. Worthen, "Life and Works of James B. Francis," in *CORHA* 5 (1894): 241.

31. G. S. Greene, J. W. Adams, and W. E. Worthen (Memoirs of Deceased Members), "James Bicheno Francis, Past President and Honorary

Member," *Proceedings of the American Society of Civil Engineers* 19 (Apr. 1893): 76.

32. Mansfield Merriman, *Treatise on Hydraulics,* 10th ed. (New York, 1916), 394.

33. John R. Freeman, "General Review of Current Practice in Water Power Production," 1924, PL&C General Files #1076.1, 22, PL&C-LNHP.

34. *Oxford English Dictionary.*

35. "Tests of Turbine Water-Wheels," *Lowell Vox Populi,* July 30, 1870.

36. Discharge diagram for "Centre Vent Water Wheel at the Boott Cotton Mills" (Mar. 1868) and table of "Quantity of Water Discharged" by a turbine at the Lawrence Manufacturing Company. Copies of both provided to the author by Al Lorenzo.

37. James Emerson, *Treatise Relating to the Testing of Water-Wheels* . . . (Willimansett, MA, 1892), 34; Thurston, "Systematic Testing of Turbine Water-Wheels," 366–67.

38. "Tests of Turbine Water-Wheels."

39. Thurston, "Systematic Testing of Turbine Water-Wheels," 367–68.

40. Ibid., 368–69, 387, 414–20.

41. *LHE* (1855), xi.

42. JBF to Washington Hunt, Jan. 20, 1858, DA5, 5, PL&C-Baker.

43. Worthen, "Life and Works of James B. Francis," 234.

44. *LHE* (1868), 169.

45. Joseph P. Frizell, *Water-Power* (New York, 1905), 527, 533–34.

46. JBF, "Gauging Water Power at Lowell, Mass.," *Engineering News,* July 25, 1878, 234–35.

47. Frizell, *Water-Power,* 532–33.

48. JBF to Benjamin Saunders, Oct. 21, 1864, DB-3, PL&C-Baker.

49. *LHE* (1868), 156.

50. JBF to Colonel Francis, Apr. 10, 1879 (#11 of European tour), Francis Family Collection, LNHP.

51. JBF to Buff & Berger, Feb. 22, 1881, JBF Records O; "Stock Account, 1885–1893," Jan. 1, 1885, both PL&C-LNHP.

52. "Experimental Comparison of Some Different Methods of Measuring the Flow of Water," *Proceedings of the Society of Arts* [at MIT], Meeting 356 (1886–87), 59.

1. Arthur Gilman, "The Freshet of 1831," *CORHA* 3 (1884–87): 270.

2. *Lowell Courier*, Jan. 12, 1841.

3. JBF, "Great Freshets in Merrimack River," *CORHA* 3 (1884–87): 252, 256.

4. Ibid., PL&C Directors, Sept. 21, 1847.

5. Sam'l M. Richardson to JBF, Feb. 23, 1850, A34, #176, PL&C-Baker; JBF, "Great Freshets," 256.

6. *Boston Daily Advertiser*, Apr. 24, 1852; "Triumph of Francis's Folly," *Boston Daily Globe*, Jan. 3, 1885.

7. JBF, "Great Freshets," 257.

8. *Boston Daily Advertiser*, Apr. 28, 1852.

9. Gilman, "The Freshet of 1831," 272.

10. Lowell Fire Department Records, 1845, Center for Lowell History, University of Massachusetts Lowell.

11. PL&C Directors, Apr. 13, 1849.

12. W C. Appleton to JBF, Dec.12, 1846; A34, #175, PL&C-Baker.

13. "Raising the Water in the Pawtucket Canal," in *Lowell Daily Journal & Courier*, July 9, 1851.

14. Ibid.

15. "Col. Francis," 1853–1864 notebook (really JBF), Dec. 15, 1852, 21, PL&C-LNHP.

16. JBF report, PL&C Directors, Feb. 21, 1848.

17. PL&C Directors, Apr. 14, 1848.

18. Ibid., Feb. 21, 1848.

19. Ibid., Apr. 8, 1857.

20. JBF to Assessors, Oct. 7, 1858, DA-5, 53, PL&C-Baker.

21. JBF sketches and notes, Apr. 1, 1848, A18, PL&C-Baker.

22. Charles Storrow to JBF, Aug. 20, 1869, in PL&C Directors.

23. George White, *Memoir of Samuel Slater*, 2nd ed. (Philadelphia, 1836), 267.

24. George Swain, "Water-power of Eastern New England," in *Reports on the Water-Power of the United States*, Vol. 16 of *Tenth Census of the United States* (Washington, DC, 1885), 83–84.

25. Caleb Kirk to E. I. DuPont, Jan. 31, 1828, quoted in Duncan Hay, "Building 'The New City on the Merrimack': The Essex Company and Its

Role in the Creation of Lawrence, Massachusetts" (Ph.D. diss., University of Delaware, 1986), 28.

26. JBF to Thomas Carey, May 17, 1858, DA-5, 18–19, PL&C-Baker.

27. JBF to Thomas Carey, Dec. 6, 1858, DA-5, PL&C-Baker.

28. PL&C Directors, Apr. 12, 1859.

29. John R. Freeman, "The Fundamental Problems of Hydroelectric Development," *Transactions of the American Society of Mechanical Engineers* 46 (1924): 529–31.

30. P. T. Jackson to Directors, Sept. 13, 1839, A1, #1; JBF to Corliss and Nightingale, Jan. 6, 1854, A2, #14, both PL&C-Baker.

31. State of Massachusetts, *Fishes, Reptiles, and Birds of Massachusetts* (Boston, 1839), 104–5.

32. Henry David Thoreau, *A Week on the Concord and Merrimack Rivers* (Mineola, NY, 2001), 20.

33. JBF Records A, Dec. 26, 1854, 197, PL&C-LNHP.

Six: Controlling the System, 1865–1885

1. PL&C Directors, Nov. 5, 1874.

2. JBF to Abbott Lawrence, July 2, 1859, A21, #110, PL&C-Baker.

3. Samuel Lawrence to Charles Hovey, Feb. 8, 1875, *CORHA* 1 (1874–79): 291.

4. *Engineering News-Record,* Mar. 29, 1928.

5. George Swain, "Water-power of Eastern New England," in *Reports on the Water-Power of the United States,* Vol. 16 of *Tenth Census of the United States* (Washington, DC, 1885), 81.

6. JBF to Thomas Carey, Dec. 6, 1859, DA-5, PL&C-Baker.

7. JBF testimony, as summarized in "Report of the Joint Special Committee . . . Passage of Fish . . . ," Apr. 1865, MA Senate Document 183, 17.

8. JBF to Charles Storrow, Apr. 21, 1866, Essex Co. Collections, MS69, Box 26, American Textile History Museum.

9. JBF Records H,10, PL&C-LNHP.

10. Ibid.

11. JBF to E. A. Bracket, clipping in Day Book, 1882–1883 (Col. Francis), June 5, 1883, PL&C-LNHP.

12. *Lowell Courier,* Feb. 20, 1869.

13. *Lowell Daily Citizen & News,* Feb. 20, 1869.

14. *Lowell Courier,* Feb. 24, 1869.

15. *Lowell Vox Populi,* Nov. 13, 1869.

16. *Lowell Courier,* Jan. 21, 1873.

17. PL&C, Directors, Feb. 24, 1885.

18. JBF to Directors, Aug. 17, 1888, in "Legal Papers" volume, PL&C library, Center for Lowell History, University of Massachusetts Lowell.

19. JBF to Charles Storrow, Oct. 27, 1867, Essex Co. Collections, Box 26, American Textile History Museum.

20. JBF to Josiah French, Aug. 30, 1858, A17, #82, PL&C-Baker.

21. William Worthen, "Life and Works of James B. Francis," *CORHA* 5 (1894): 235.

22. PL&C Directors, Mar. 3, 1870.

23. JBF annual report, Sept. 21, 1869. PL&C-Baker.

24. James Francis [Colonel], "Biographical Sketch of the Life of Cleveland J. Cheney, *CORHA* 6 (1896–1904): 114–19.

25. JBF Records H, 131, PL&C-LHNP.

26. JBF to Directors, Aug. 17, 1888, in "Legal Papers."

27. Hiram Mills, "The Proprietors of Locks and Canals on Merrimack River" (1907), 3, Center for Lowell History, University of Massachusetts Lowell.

28. JBF to PL&C "Committee on Deepening Hunt's Falls," DB-8, Feb. 1, 1876, PL&C-Baker; Worthen, "Life and Works of James B. Francis," 237.

29. PL&C Directors, May 28, 1849 and Dec. 30, 1850; Paul Hill, "Personal Reminiscences of Lowell, Fifty Years Ago," *CORHA* 5 (1894): 293.

30. JBF to Col. Francis, Jan. 24, Feb. 3, and May 5, 1879, Francis Family Collection, LNHP.

31. Directors' Records, Feb. 2, 1839.

32. JBF to Col. Francis, May 19 and Apr. 10, 1879, Francis Family Collection, LNHP.

33. JBF, "Address," *Transactions of the American Society of Civil Engineers* 10 (1881): 190.

34. JBF to Col. Francis, Feb. 3, 1879, Francis Family Collection, LNHP.

35. Ibid., Feb. 16, 1879.

36. "Moving a Large Tree," *Lowell Vox Populi,* July 24, 1875.

37. Worthen, "Life and Works of James B. Francis," 238.

38. PL&C Directors, Feb. 27, 1879.

39. Hiram Mills, "James Bicheno Francis," for MIT (Cambridge, MA, 1892), 6–7; Worthen, "Life and Works of James B. Francis," 230, 238.

40. *Chronicle and Constitutionalist* (Augusta, GA), Mar. 23, 1884.

41. PL&C Directors, Nov. 20, 1884.

Postscript

1. William Worthen, "Life and Works of James B. Francis," *CORHA* 5 (1894): 240.

2. Ibid., 241.

3. Joseph P. Frizell, "Reminiscences of James B. Francis," *Engineering News,* July 12, 1894, 29.

4. Ibid., 30; Desmond Fitzgerald, Joseph P. Davis, and John R. Freeman, "James Bicheno Francis: A Memoir," *Journal of the Association of Engineering Societies* 13 (Jan. 1894), 9.

5. Laurence Gross, *The Course of Industrial Decline: The Boott Cotton Mills of Lowell, Massachusetts, 1835–1955* (Baltimore, 1993), 94–95.

6. J. A. Hunnewell to M. E. Sperry, Stone & Webster, Inc., Oct. 3, 1927. Letter copy from Charles Parrott, LNHP.

7. Alan Steiner, "The Electrification of Mills on the Lowell Canal System," research paper in a Brown University seminar taught by Patrick Malone, ca. 1980.

8. Personal Communications with Mel Lezberg, Dec. 18, 2007, and Jan. 10, 2008.

9. Steve Turner and Charles Scullion Jr., *Working the Water* (Lowell: Lowell Historic Preservation Commission, n.d.).

Suggested Further Readings

....

There is a rich literature on Lowell, industrialization, and waterpower. This brief essay suggests books that are relevant and accessible for readers of *Waterpower in Lowell*. Anyone who intends to do research with the primary sources on Lowell should visit the Center for Lowell History of the University of Massachusetts Lowell, the Baker Library Historical Collections of Harvard Business School, the Massachusetts Historical Society, the Lowell National Historical Park, and the American Textile History Museum.

For understanding the historical context and impact of industrialization in the United States, several American history textbooks are excellent starting points. Pauline Maier, Merritt Roe Smith, and Alexander Keyssar's *Inventing America* (New York, 2005) is sophisticated in its approach to industry and the history of technology. John Farragher, Mari Jo Buhle, Daniel Czitrom, and Susan Armitage's *Out of Many*, vol. 1 (Upper Saddle River, NJ, 2006) uses Lowell and its labor history as an important case study. A wealth of drawings and maps makes the National Geographic Society's *Historical Atlas of the United States* (Washington, DC, 1989) an ideal complement to either of those textbooks.

Ruth Schwartz Cowan's *A Social History of American Technology* (New York, 1997), Alan Marcus and Howard Segal's *Technology in America* (Orlando, FL, 1999), and Carroll Pursell's *The Machine in America: A Social History of Technology* (Baltimore, 2007) survey the American experience with technology and give nuanced interpretations of the effects of mechanization. Thoughtful essays and an erudite evaluation of books on technological history distinguish *Early American Technology: Making and Doing Things from the Colonial Era to 1850* (Chapel Hill, NC, 1994), edited by Judith McGaw.

Eugene Ferguson's *Engineering and the Mind's Eye* (Cambridge, MA, 1993) focuses on the ways engineers think and how they make use of drawings and models in the design process. His student David Hounshell offers astute insights about manufacturing in *From the American System to Mass Production* (Baltimore, 1984), which includes an "artifactual analysis" of Singer sewing machines.

Physical evidence can tell us a great deal about the processes and effects of industrialization. Brooke Hindle and Steven Lubar wrote *Engines of Change: The Industrial Revolution in America, 1790–1860* (Washington, DC, 1986) to complement their highly successful museum exhibition at the Smithsonian Institution. In *The Texture of Industry: An Archaeological View of the Industrialization of North America* (New York, 1995), Robert B. Gordon and Patrick Malone use the techniques of industrial archaeologists to investigate power systems, landscapes, manufacturing processes, and work in shops and factories. David Meyer applies his expertise in historical geography to explore *The Roots of American Industrialization* (Baltimore, 2003).

David Nye's *American Technological Sublime* (Cambridge, MA, 1996) and *America as Second Creation: Technology and Narratives of New Beginnings* (Cambridge, MA, 1996) are interdisciplinary studies of American attitudes about technology and industrialization. Equally relevant is his *Consuming Power: A Social History of American Energies* (Cambridge, MA, 1999). Nye is one of many notable scholars strongly influenced by Leo Marx's seminal work, *Machine in the Garden: Technology and the Pastoral Ideal in America* (New York, 2000), which describes the reactions of literary figures like Thoreau and Emerson to waterpower development in the Merrimack River Valley. See also Merritt Roe Smith and Leo Marx's edited volume *Does Technology Drive History?* (Cambridge, MA, 1994), which tackles the thorny issue of technological determinism.

Environmental history has grown rapidly as an academic field in recent decades and is now closely linked to the history of technology and urban history. One of the best books in that field is about human interactions with the Merrimack River: Theodore Steinberg's *Nature Incorporated: Industrialization and the Waters of New England* (Cambridge, 2004) offers a critical perspective on the impacts of industrialization at Lowell and the creation of the reservoir system in New Hampshire. Diana Muir's *Reflections in Bullough's Pond: Economy and Ecosystem in New England* (Hanover, NH, 2000) provides a regional overview of economic development and how it altered natural ecology. Chad Montrie's *Making a Living: Work and Environment in the United States* (Chapel Hill, NC, 2008) shows how much Lowell "mill girls" appreciated both natural scenery and the plantings on corporate property.

For an impressive historiographical essay on works dealing with the connections between technology and the environment, see Joel Tarr and Jeffrey Stine, "At the Intersection of Histories: Technology and the Environment," *Technology and Culture* 39.4 (1993). Tarr, the dean of American environmental historians, recently published *Devastation and Renewal: An Environmental History of Pittsburgh* (2005), with contributions by him and other scholars. The chapter by Edward K. Muller, *"River City,"* is particularly relevant for readers of this volume. Another essential book for anyone studying the way our technological choices affect rivers is Richard White's *The Organic River: The Remaking of the Columbia River* (New York, 1996). Worthy of special note is White's discussion of waterfalls and rapids as "critical sites in a geography of energy."

Lowell once depended heavily on and still makes significant use of energy from the Pawtucket Falls of the Merrimack River. The best introduction to Lowell, and a model for the interpretation of national historical parks, is *Lowell: The Story of an Industrial City* (Washington, DC, 1992). That official handbook was written by Thomas Dublin, winner of the Bancroft Prize for *Women at Work: The Transformation of Work and Community in Lowell, Massachusetts, 1826–1860* (New York, 1993). Mary Blewett's introduction to *Caught between Two Worlds: The Diary of a Lowell Mill Girl* (Lowell, MA, 1984) comments on the experiences of Susan Brown in a water-powered Lowell mill. Steve Dunwell's *Run of the Mill* (Boston, 1978) is both a history of the New England textile industry and a gallery of superb illustrations. There is much on textiles and on Lowell in Paul Hudon's *Lower Merrimack: The Valley and Its Peoples* (Sun Valley, CA, 2004) and in Carl Siracusa's *A Mechanical People: Perceptions of the Industrial Order in Massachusetts, 1815–1880* (Middletown, CT, 1979).

Robert Weible, formerly the historian of the Lowell National Historical Park, has included waterpower papers by Patrick Malone and by Charles and Pauline Carroll in his edited compilation *The Continuing Revolution: A History of Lowell, Massachusetts* (Lowell, MA, 1991). The Carrolls astutely present Francis's empirical research as an example of scientific engineering. The same volume has many other relevant contributions, including Weible's original research on East Chelmsford before it became Lowell, Lawrence Gall's analysis in "The Heritage Factor in Lowell's Revitalization," Betsy Peterson's exploration of textile mill design, and Laurence Gross's conclusions about labor and management at the Boott Cotton Mills, which he covers in more detail in *The Course of Industrial Decline: The Boott Cotton Mills of Lowell, Massachusetts, 1835–1955* (Baltimore, 2000). An earlier anthology, *Cotton Was King: A History of Lowell, Massachusetts* (Lowell, MA, 1976), edited by Arthur L. Eno Jr., remains helpful.

George Gibb makes the Boston Manufacturing Company and the Proprietors of Locks and Canals key parts of his great business history, *The Saco-Lowell Shops* (Cambridge, MA, 1959). Heidi Vernon-Wortzel's *Lowell: The Corporations and the City* (New York: 1992) discusses company involvement in urban affairs. The investors and corporate leaders known as the Boston Associates are the subjects of Robert Dalzell's influential *Enterprising Elite: The Boston Associates and the World They Made* (New York, 1993). Frances W. Gregory applies her formidable research skills to one of the associates who founded Lowell in *Nathan Appleton: Merchant and Entrepreneur, 1779–1861* (Charlottesville, VA, 1988). Brad Parker's *Kirk Boott* (Lowell, MA, 1985) gives a biographical sketch of another founder.

Brian Mitchell's *The Paddy Camps: The Irish of Lowell, 1821–*

1861 (Urbana, IL, 2006) adds details on members of one immigrant group that helped build and maintain Lowell's canals, dams, and industrial buildings. Steve Turner and Charles Scullion Jr.'s *Working the Water* (Lowell, MA, n.d.), a booklet from the Lowell Historical Preservation Commission, is the only publication devoted to the workers who operated the complex Locks & Canals system. *The Lower Merrimack River Valley* (North Andover, MA, 1978), edited by Peter Molloy, contains site descriptions by Charles Hyde and Charles Parrott from the 1974–75 Historic American Engineering Record (HAER) survey of the Lowell Canal System.

Waterpower as a key resource for industrialization is covered very well in Terry Reynolds's *Stronger Than a Hundred Men* (Baltimore, 1983), Richard Hills's *Power in the Industrial Revolution* (Manchester, UK, 1970), Lucille Kane's *The Falls of St. Anthony: The Waterfall that Built Minneapolis* (St. Paul, MN, 1987), and Louis Hunter's *Waterpower in the Century of the Steam Engine* (Charlottesville, VA, 1979). Hunter's classic study, which gives much attention to Lowell, James B. Francis, and water turbines, was the first volume of his magisterial series *A History of Industrial Power in the United States, 1780–1930*. Volume 2, *Steam Power* (Charlottesville, VA, 1986), includes comparisons of steam and waterpower. Volume 3, *The Transmission of Power* (Cambridge, MA, 1991), which was completed by Lynwood Bryant, has details on the gearing, belting, and shafting used in textile mills.

Both Terry Reynolds and Richard Hills contributed thought-provoking papers for *The World of the Industrial Revolution: Comparative and International Aspects of Industrialization* (North Andover, MA, 1986), edited by Robert Weible. Reynolds is an accomplished historian of technology who has looked closely at the testing of vertical waterwheels and at the adoption of the breast wheel. The foremost expert on turbine development and flow measurement is Edwin Layton, whose extensive publication record includes a perceptive chapter on engineer James B. Francis in *Technology in America: A History of Individuals and Ideas* (Cambridge, MA, 1990), edited by Carroll Pursell. Layton also wrote a remarkable book encouraging quantitative reasoning in liberal education, *From Rule of Thumb to Scientific Engineering:*

James B. Francis and the Invention of the Francis Turbine (Stony Brook, NY, 1992). Hunter Rouse gives an overview of the hydraulic engineering profession that Francis helped to shape in *Hydraulics in the United States, 1776–1976* (Iowa City, IA, 1976).

Jeffrey F. Mount's well-illustrated *California Rivers and Streams* (Berkeley, CA, 1995) explains principles of hydrology and geomorphology. Hydrological historian Martin Reuss introduces a theme issue, *Technology and Culture* 49, no. 3 (2008), with articles on water projects throughout the world. Robert B. Gordon's "Hydrological Science and the Development of Waterpower for Manufacturing," *Technology and Culture* 26, no. 2 (1985) shows how reservoirs improved flow in dry seasons and demonstrates the value of hydrological analysis in historical evaluations of waterpower potential at New England sites.

Elizabeth Sharpe, like Gordon, stresses the importance of reservoirs for reliable waterpower, but her book, *In the Shadow of the Dam: The Aftermath of the Mill River Flood of 1874* (New York, 2004) is about the tragic collapse of a poorly designed dam in Massachusetts. James B. Francis was one of the engineers who was asked to assess the causes of that disaster. Donald C. Jackson's *Great American Bridges and Dams* (Washington, DC, 1988) assesses a broad spectrum of dams while deftly explaining their historical context and design principles. For understanding how milldams, waterwheels, turbines, and power transmission systems worked, nothing is better than author/illustrator David Macaulay's *Mill* (Boston, 1989), which is not just for younger readers.

Urban canals, used for both power and transportation, are the focus of *Canals and American Cities* (Easton, PA, 1993). The annual volumes of *Canal History and Technology Proceedings* (Easton, PA, 1982–2008), edited by Lance Metz, continue to present stimulating scholarship on canals of all types. Christopher Roberts tells the history of the pioneering waterway that connected the Merrimack River to Boston in *The Middlesex Canal, 1793–1860* (Cambridge, MA, 1938).

A number of books look at the design of manufacturing towns, but the landscaping of company property, including spaces around water features, deserves more attention. Thomas Bender, in *Toward an Urban*

Vision: Ideas and Institutions in Nineteenth-Century America (Baltimore, 1982), uses paintings as well as documentary evidence in his investigations of Lowell's urban form and the debates about public space in the growing city. Alan Emmet's *So Fine a Prospect: Historic New England Gardens* (Hanover, NH, 1996) reveals the deep horticultural and botanical roots of the Boott family. John Coolidge's pioneering volume on Lowell's architecture and urban development, *Mill and Mansion* (Amherst, MA, 1993), still has much to offer, particularly on ideal layouts for mills, canals, and company housing. Stephen Mrozowsky, Grace Zeisling, and Mary Beaudry skillfully apply the methods of landscape archaeology in *Living on the Boott: Historical Archaeology at the Boott Mills Boardinghouses, Lowell, Massachusetts* (Amherst, MA, 1996). Margaret T. Parker's *Lowell: A Study of Industrial Development* (New York, 1940) provides very useful information on the spatial patterns of mill placement in this city. Lowell receives some attention in Margaret Crawford's *Building the Workingman's Paradise: The Design of American Company Towns* (New York, 1995), but the book does more with industrial communities that came later, including cotton mill towns of the American South.

Historical studies of urban planning at other manufacturing sites are valuable for comparative purposes: see Richard Candee's *Atlantic Heights* (Portsmouth, NH, 1985), John Garner's *The Model Company Town* (Amherst, MA, 1984), Garner's edited volume, *The Company Town* (New York, 1992), and John Armstrong's *Factory under the Elms: A History of Harrisville, New Hampshire* (Cambridge, MA, 1970). Kingston Heath has a fine chapter on the design of New Bedford's Howland Mill Village in his *Patina of Place* (Knoxville, TN, 2001). *Green Engineering: Parks and Promenades in the Industrial Community*, a special theme issue of *IA: The Journal of the Society for Industrial Archeology* 24.1(1998), edited by Patrick Malone, has an article on Lowell by Malone and Charles Parrott, on Georgiaville and Providence, Rhode Island, by Richard Greenwood, and on Calumet and Gwinn, Michigan, by Arnold Alanen and Lynn Bjorkman.

Alanen is the co-editor, with Robert Melnick, of the influential volume *Preserving Cultural Landscapes in America* (Baltimore, 2000).

Preserving industrial landscapes like the Lowell Canal System and the mill yards it served is part of the effort to save and interpret places that matter. Several provocative writers have taken a strong stand on the cultural conservation of place: see Dolores Hayden, *The Power of Place: Urban Landscapes as Public History* (Cambridge, MA, 1997), and the chapter by Setha Low in *Conserving Culture*, ed. Mary Hufford (Urbana, IL, 1994). National Park Service architect Charles Parrott, author of *Lowell Then and Now: Restoring the Legacy of a Mill City* (Lowell, MA, 1995), has been a consistent advocate for rehabilitation and re-creation of buildings and landscapes in Lowell.

Lowell is often seen as a striking example of urban revitalization through historical preservation, public history programming, and industrial heritage tourism. *New Perspectives on Industrial History Museums*, a special theme issue edited by Stephen Cutcliffe and Steven Lubar for the *Public Historian* 22.3 (2000), includes fascinating commentary on exhibits in Lowell. The Lowell National Historical Park, which interprets waterpower and other subjects for thousands of visitors each year, gets extended treatment in Kathy Stanton's *The Lowell Experiment: Public History in a Postindustrial City* (Amherst, MA, 2006). Martha Norkunas's *Monuments and Memory: History and Representation in Lowell, Massachusetts* (Washington, DC, 2002) examines the creation of memorials in the city, including a prominent fountain with large granite blocks and a life-size canal worker in bronze.

Index

....

Page numbers in *italics* refer to illustrations.

canal construction, with draft animals, 28, 30

canal locks: collapse of, 12–13; operation of, 11–12, 31. *See also* locks on Lowell Canal System

canals, transportation: Erie, 28; Great Dismal Swamp, 12; Middlesex, 10, 13, 15, 22, 51, 70–71. *See also* Pawtucket Canal

canal systems: island concept in, 5, 25, 36, *38–39*, 169; with multiple levels, 4, 31, *35*, 169, 170. *See also* Lowell Canal System, two levels of

census, tenth U.S. (1880), 1, 187

Cheney, Cleveland, 203

Chicopee, MA, 168

city formation at waterfalls, 168–69, 219

Civil War, and Lowell, 178–179, 181

Clark, Thomas, 23

Cohoes, NY, 170, 216

Colburn, Warren, 56

Colt, Peter, 4

Columbus, GA, 170

commodification of water, 41–42, 183–84, 221–22. *See also* millpowers; surplus water

concrete, 97–98

contractors, canal system, 28–29, 89, 92

Corliss, George, 174, 190

corporations, textile, 1, 2, 4, 90. *See also* textile production; *and names of individual corporations*

Cummiskey, Hugh, 28–29

dams, 4, 8, 96; failure of 169, 205, 216. *See also* fishways; flashboards; Pawtucket Dam

depressions, economic 12, 62 , 178

Dover, NH, 168

Dummer, John, 32, 55, 63, 103–4, 113

dynamometer (brake), 56, 111–12, 129

East Chelmsford, 9, 11, 18–20, 21–22; land acquisition in, 23–24, 43–44; topography of, *14*, 25. *See also* Pawtucket Falls

Eastern Canal, 37, *54*, 61

efficiency: of breastwheels, 40, 56, 65, 104; of turbines, 40, 107, 109, 112–13, 115, 119

Emerson, James, 129–32; as author, 132

Emerson, Ralph Waldo, 86

Essex Company, 75, 83, 85–86, 96, 166–67, 192

explosives: blasting with, 30, 92–94, 99; from gunpowder mills, 19, 93, 99

factory system, 57

Fall River, MA, 175, 190

fire protection, 150–53, 196, 224

fish and fishermen, 9, 45–47, 176–77, 192–94

fishways, 48, 177, 192–94

flashboards, 64, 95, 180, 205–8

floods, 96–97, *97*, 143–46, *149*, 216, 223

flowage payments, 64, 86

flow measurement, 101, 222; with apertures, 170; with box capture, 101, 125; with current meters, 139–42; with gauge wheels, 56, 102–4; with rise in millpond, 68; with speed gate settings, 129–30, 136–37, 170; with surface floats, 102–4; with tubes, 134–42, *138*, 170, 173, 224; with weirs, 110, 124, 126, 128, 134, 170, 202

flumes: in canals, 102–4, 128, 134–42, *138*, 172–73; testing (for turbines), 128–33

Francis, James (Col.), 173, 214–15, 217

Francis, James B., 6, 220–21; accuracy of, 173, 205; as agent, 75, 82, 194, 221; and American Society of Civil Engineers, 60, 137, *138*, 205,